Experiments and Numerical Simulations of Diluted Spray Turbulent Combustion

ERCOFTAC SERIES

VOLUME 17

Aims and Scope of the Series

ERCOFTAC (European Research Community on Flow, Turbulence and Combustion) was founded as an international association with scientific objectives in 1988. ERCOFTAC strongly promotes joint efforts of European research institutes and industries that are active in the field of flow, turbulence and combustion, in order to enhance the exchange of technical and scientific information on fundamental and applied research and design. Each year, ERCOFTAC organizes several meetings in the form of workshops, conferences and summerschools, where ERCOFTAC members and other researchers meet and exchange information.

The ERCOFTAC Series will publish the proceedings of ERCOFTAC meetings, which cover all aspects of fluid mechanics. The series will comprise proceedings of conferences and workshops, and of textbooks presenting the material taught at summerschools.

The series covers the entire domain of fluid mechanics, which includes physical modelling, computational fluid dynamics including grid generation and turbulence modelling, measuring-techniques, flow visualization as applied to industrial flows, aerodynamics, combustion, geophysical and environmental flows, hydraulics, multiphase flows, non-Newtonian flows, astrophysical flows, laminar, turbulent and transitional flows.

For further volumes:
http://www.springer.com/series/5934

Bart Merci · Dirk Roekaerts · Amsini Sadiki
Editors

Experiments and Numerical Simulations of Diluted Spray Turbulent Combustion

Proceedings of the 1st International Workshop on Turbulent Spray Combustion

Springer

Editors
Bart Merci
Flow, Heat and Combustion Mechanics
Ghent University
Sint-Pietersnieuwstraat 41
B-9000 Ghent
Belgium
Bart.Merci@UGent.be

Dirk Roekaerts
Department of Multi-Scale Physics
Delft University of Technology
PO Box 5
2600 AA Delft
Netherlands
d.j.e.m.roekaerts@tudelft.nl

Amsini Sadiki
Department of Mechanical Engineering
Darmstadt University of Technology
Petersenstrasse 30
64287 Darmstadt
Germany
sadiki@ekt.tu-darmstadt.de

ISSN 1382-4309
ISBN 978-94-007-1408-3 e-ISBN 978-94-007-1409-0
DOI 10.1007/978-94-007-1409-0
Springer Dordrecht Heidelberg London New York

Library of Congress Control Number: 2011929035

Cover design: eStudio Calamar S.L.

Printed on acid-free paper

Springer is part of Springer Science+Business Media (www.springer.com)

Preface

This book reflects the outcome of the 1st International Workshop on Turbulent Spray Combustion held in 2009 in Corsica (France). The focus is on reporting the progress of experimental and numerical techniques in two-phase flows, with emphasis on spray combustion. The motivation for studies in this area is that knowledge of the dominant phenomena and their interactions in such flow systems is essential for the development of predictive models and their use in combustor and gas turbine design. This necessitates the development of accurate experimental methods and numerical modelling techniques. The workshop aimed at providing an opportunity for experts and young researchers to present the state-of-the-art, discuss new developments or techniques and exchange ideas in the areas of experimentations, modelling and simulation of reactive multiphase flows.

The first two chapters reflect the contents of the invited lectures, given by experts in the field of turbulent spray combustion. The first chapter concerns computational issues, while the second deals with experiments. These lectures initiated very interesting and interactive discussions among the researchers, further pursued in contributed poster presentations. Chapters 3 and 4 focus on some aspects of the impact of the interaction between fuel evaporation and combustion on spray combustion in the context of gas turbines, while the final chapter deals with the interaction between evaporation and turbulence. In short, we believe that these contributions individually and together are highly valuable for the research community. They allow experts and young scientists to get insight in the many aspects of dilute turbulent spray combustion.

It is our intention to continue our efforts to bring together experimentalists and modellers of spray combustion. As a next step, in preparation of subsequent workshops, we intend to define 'target test cases', for which experimental databases are available or will be made available and on which modelling and numerical algorithm issues will be tested. We hope you enjoy reading this book and that it may be the onset for you to participate in this quest.

Bart Merci
Dirk Roekaerts
Amsini Sadiki

The kind support of ERCOFTAC and SFB568 (TU-Darmstadt) to the organisation of the 1st Workshop is strongly appreciated.

Contents

Contributors

W. Ahmadi Department of Mechanical and Processing Engineering, Institute for Energy and Powerplant Technology, Technische Universität Darmstadt, Petersenstr. 30, 64287 Darmstadt, Germany, ahmadi@ekt.tu-darmstadt.de

Nijso A. Beishuizen NUMECA International, Development Engineer, Multiphysics CFD Group, Ch. de la Hulpe 189 Terhulpse Steenweg – 1170 Brussels – Belgium, http://www.numeca.be

Mouldi Chrigui Department of Mechanical and Processing Engineering, Institute of Energy and Power Plant Technology, Darmstadt University of Technology, Petersenstr. 30, D-64287 Darmstadt, Germany, mchrigui@ekt.tu-darmstadt.de

James D. Gounder School of Aerospace, Mechanical and Mechatronic Engineering, The University of Sydney, NSW, 2006 Australia
and
Deutsches Zentrum fr Luft- und Raumfahrt in der Helmholtz-Gemeinschaft Institut für Verbrennungstechnik, 70569 Stuttgart Germany, James.Gounder@dlr.de

Eva Gutheil Interdisziplinäres Zentrum für Wissenschaftliches Rechnen, Ruprecht-Karls-Universität Heidelberg, Im Neuenheimer Feld 368, 69120 Heidelberg, Germany, gutheil@iwr.uni-heidelberg.de

J. Janicka Department of Mechanical and Processing Engineering, Institute for Energy and Powerplant Technology, Technische Universität Darmstadt, Petersenstr. 30, 64287 Darmstadt, Germany
and
Center of Smart Interface, Technische Universität Darmstadt, Petersenstr. 32, 64287 Darmstadt, Germany, janicka@ekt.tu-darmstdt.de

Assaad R. Masri School of Aerospace, Mechanical and Mechatronic Engineering, The University of Sydney, NSW, 2006 Australia, assaad.masri@sydney.edu.au

Dirk Roekaerts Department of Multi-Scale Physics, Delft University of Technology, PO Box 5, 2600 AA Delft, The Netherlands, d.j.e.m.roekaerts@tudelft.nl

Amsini Sadiki Department of Mechanical and Processing Engineering, Institute for Energy and Powerplant Technology, Technische Universität Darmstadt, Petersenstr. 30, 64287 Darmstadt, Germany
and
Center of Smart Interface, Technische Universität Darmstadt, Petersenstr. 32, 64287 Darmstadt, Germany, sadiki@ekt.tu-darmstadt.de

Chapter 1
Issues in Computational Studies of Turbulent Spray Combustion

Eva Gutheil

Abstract Turbulent spray combustion occurs in many technical applications such as internal engine combustion, gas turbine combustion, liquid-fueled rockets, and industrial burners. Major challenges are the modeling of detailed processes including the atomization process, the turbulent flow field, particle motion and interaction, chemical reactions as well as the strong coupling between these processes. The method of choice to achieve an integral model for the modeling and simulation of turbulent spray combustion is the detailed modeling of fundamental processes and simplification of these models before they enter a more complex tool. Thus, it is guaranteed that models are based on physical grounds and the degree of detailedness is sufficient to capture the essential features of the underlying process. The final tool then is based on physical grounds and it is not burdened with details not contributing to the main features of the flame structure.

The present work aims to present an overview on the state of the art with respect to underlying physical models and challenges that need to be addressed in the near future. They include the modeling and simulation of turbulence, turbulent mixing as well as the interaction of evaporation with the turbulent flow field and the chemical reactions. Principal approaches to modeling sprays in a turbulent flow field such as RANS (Reynolds averaged Navier-Stokes equations), DNS (direct numerical simulation) and LES (large eddy simulation) are addressed. Moreover, Reynolds stress models (RSM), PDF (probability density function), and CMC (conditional moment closure) models will be presented and discussed. These approaches can act both as general spray models and as subgrid models for LES.

In order to address environmental concerns, detailed chemical kinetics need to be considered to account for the prediction of pollutant emission and its reduction. Appropriate methods such as direct closure methods and (spray) flamelet models for turbulent spray flames are addressed. Also, reduced chemical mechanisms are discussed.

E. Gutheil (✉)
Interdisziplinäres Zentrum für Wissenschaftliches Rechnen, Ruprecht-Karls-Universität Heidelberg, Im Neuenheimer Feld 368, 69120 Heidelberg, Germany
e-mail: gutheil@iwr.uni-heidelberg.de

B. Merci et al. (eds.), *Experiments and Numerical Simulations of Diluted Spray Turbulent Combustion*, ERCOFTAC Series 17, DOI 10.1007/978-94-007-1409-0_1,

It is shown that most often gas flame models are not appropriate for spray flame modeling because of the high impact that both spray dynamics and evaporation have on spray combustion.

Moreover, numerical results using different models and experimental data mainly of research burners are discussed. The scope is the evaluation of present models, and future research areas of interest are identified.

Keywords Spray combustion · Modeling and simulation · Two-phase flows · Reactive flows

1 Introduction

Liquid fuels are used in many practical combustion systems such as internal combustion engines, gas turbines, liquid-fueled rockets, and they significantly contribute to our energy supply. Liquid fuels typically are delivered into the combustion chamber as turbulent sprays. The combustion efficiency, combustion stability, and pollutant formation strongly depend on the character of the turbulent spray, and an improved understanding of the turbulent spray combustion systems is required. Compared to gas phase combustion, the two-way coupling of the liquid and the gas is an additional challenge in these systems. The modeling and simulation of turbulent spray combustion is very complex because these processes involve turbulence, heat and mass transfer, phase change, and chemical reactions. Moreover, all underlying processes are strongly coupled. The reliable prediction of pollutant emissions asks for the consideration of detailed chemistry where non-equilibrium effects need to be taken into account.

Modeling and simulation of technical sprays are essential tools for the design of technical applications, and the study of their performance, efficiency, and optimization are vital to both spray flows and flames. However, a numerical tool to design the complete combustion process is not available, and current studies include derivation and evaluation of models, their numerical solution and comparison with experiment. In spray combustion systems, most often the process of injection and atomization is neglected due to its complexity and the lack of suitable models, and very often, initial profiles of droplet distributions are taken from experiments after atomization.

Droplet evaporation and combustion are the underlying processes of any technical spray application, and special focus must be given to these processes. In particular, droplets interacting with the (laminar or turbulent) flow field and the flame as well as droplets influencing each other or droplets which interact with walls and films are of particular interest in spray systems.

The most essential processes in a spray flame include injection and atomization, droplet evaporation, consideration of the separated flow and the gas phase including turbulence and chemical reactions. They strongly interact with each other and need to be considered in turbulent spray flame computations. The challenge of numerical simulations, in general, is to find a compromise between computational efficiency including the simplification of mathematical models and the need for detailed mathematical description of relevant processes determining the entire performance.

In this situation, it is advisable to build models for underlying processes, evaluate and simplify these models in order to include them in more complex applications and simulations. Thus, the present survey concerns the mathematical base of droplet and spray evaporation and combustion. In particular, single droplets, interacting droplets, and sprays are considered in convective both non-reacting and reacting environments. Essentials here are the coupling of droplet evaporation and turbulence as well as chemical reactions and the description of the spray. The latter process includes droplet interaction, breakup and collisions as well as agglomeration.

There are several methods to account for the consideration of the droplets in a dilute spray. They include models such as the discrete droplet model (also known as particle-source-in-cell model), population models as well as methods of moments. Major features, advantages and disadvantages are discussed.

Moreover, focus is on the modeling and simulation of turbulence, turbulent mixing as well as the interaction of evaporation with the turbulent flow field and the chemistry. Principal approaches to modeling sprays in a turbulent flow field such as RANS (Reynolds averaged Navier-Stokes equations), DNS (direct numerical simulation) and LES (large eddy simulation) are addressed. Also, Reynolds stress models (RSM), PDF (probability density function), and CMC (conditional moment closure) models are presented and discussed.

In the case of spray flames, chemical reactions must be considered. In order to address environmental concerns, detailed chemical kinetics must be included to account for the prediction of pollutant emission and its reduction. Among others, direct closure methods and (spray) flamelet models for turbulent spray flames are addressed. Also, reduced chemical mechanisms are considered which consume less computational time and thus are more appropriate for technical applications. However, their prediction of pollutants is less powerful and fails in situations where intermediates deviate from chemical equilibrium and steady state away from the main chemical reaction zone.

In areas where both chemical reactions and spray evaporation occur, standard models transferred from gas-phase models fail to predict major features of spray flames. These regimes are particularly challenging since in these areas the coupling of evaporation and combustion is strongest, and special care is required.

The survey shows a comparison of numerical results using different models and experimental data mainly of research burners to estimate the capacities and shortcomings of present models, and future research areas of interest are identified.

2 Spray Atomization

Spray injection and atomization are not well understood even though they are the processes which dominate the entire spray characteristics. Most modeling studies start with the atomized spray, and initial droplet distributions from experimental data are taken to define the inlet conditions for the simulation. The dense spray region is difficult to access both with experimental and numerical models. Laser-optical methods cannot penetrate the dense spray and thus, characterization of this regime is challenging. Numerical approaches include the dense spray core as a separate phase, and a transitional region surrounding this core is described to account for the atomization regime.

Numerical models for spray combustion range from one-dimensional models to two- and three-dimensional models, where steadiness or time-dependence may be chosen depending on the application. Figure 1.1 gives an overview of standard

Application	Configuration	Independent Variables[b]	Structure
Prevaporizing system: afterburners, lean combustors, carburetors, ramjets	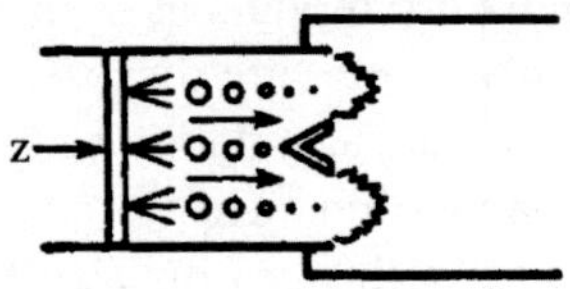	z	Steady nonburning
Liquid-fueled rocket engines	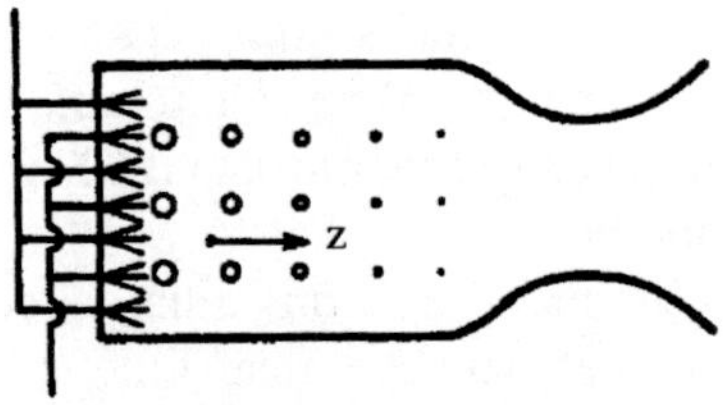	z	Steady, more or less premixed
Gas-turbine combustors	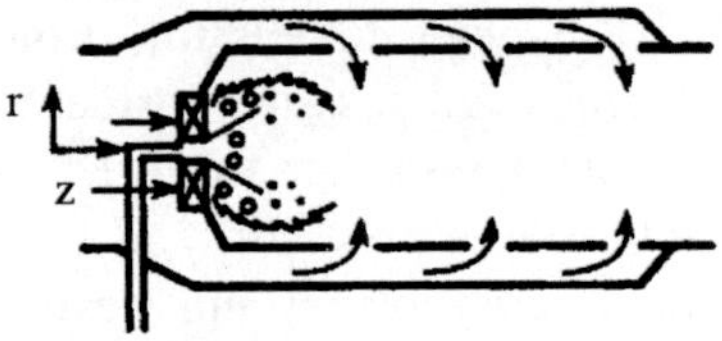	z	Steady, diffusion flame
Industrial furnaces	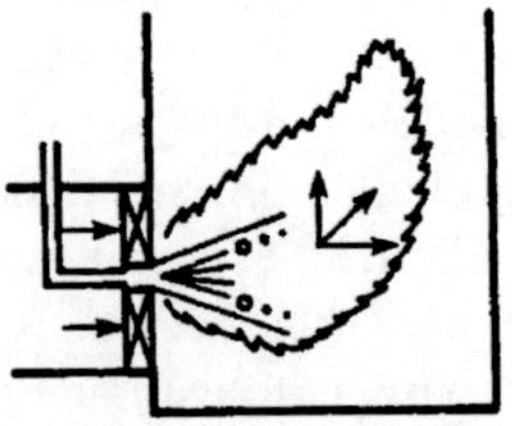	x,y,z	Steady, diffusion flame
Diesel engines		t,x,y,z	Transient, diffusion flame, ignition characteristics needed

Fig. 1.1 Spray combustion systems (Reprinted with permission from Ref. [1]; © 1977, Elsevier)

applications and the minimum degree of effort in terms of dimension and time dependence [1]. Increased computer power enables the simulation of the three-dimensional unsteady configuration, for instance, in Diesel engine combustion. However, detailed models for all processes cannot be incorporated in such complex simulations because of long and costly computational times.

Figure 1.2 displays different types of pressure and twin fluid atomizers [1]. Atomizers are characterized for different applications and the resulting spray structure is given. In particular, the formation of hollow and full cone or multiple cone structures strongly affect spray dynamics. With respect to modeling issues, the full cone spray is easier to access because it has only one interfacial boundary where atomization and interaction with the surrounding gas phase may occur.

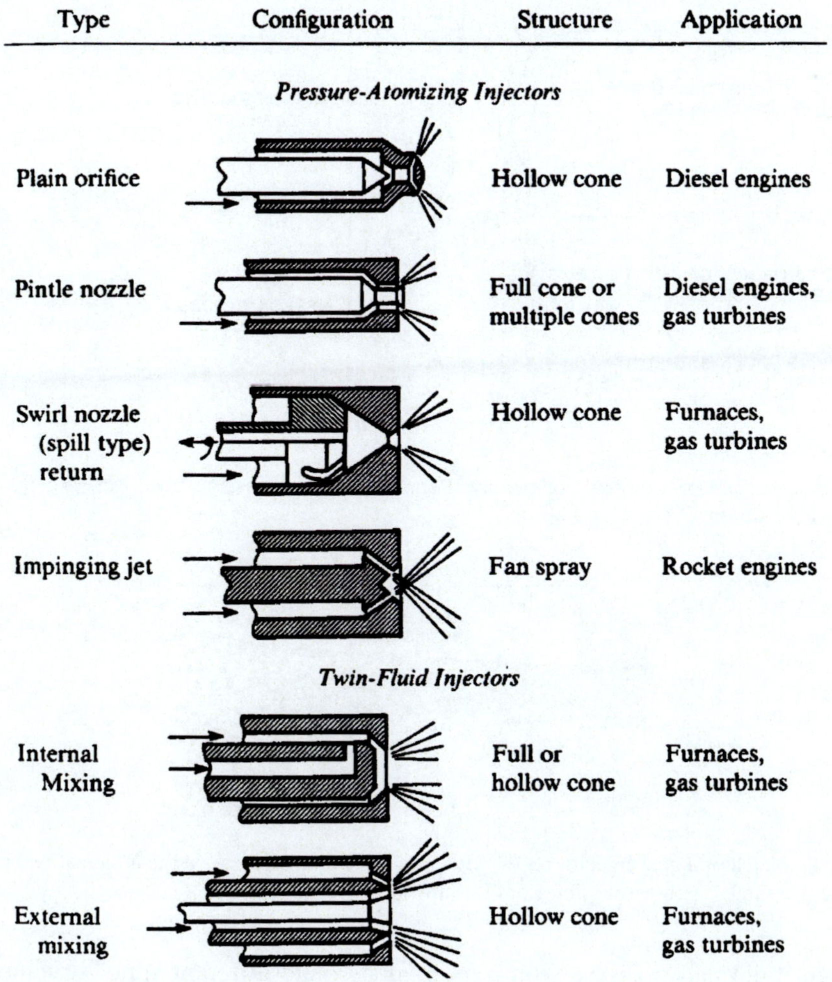

Fig. 1.2 Atomizers and their applications (Reprinted with permission from Ref. [1]; © 1977, Elsevier)

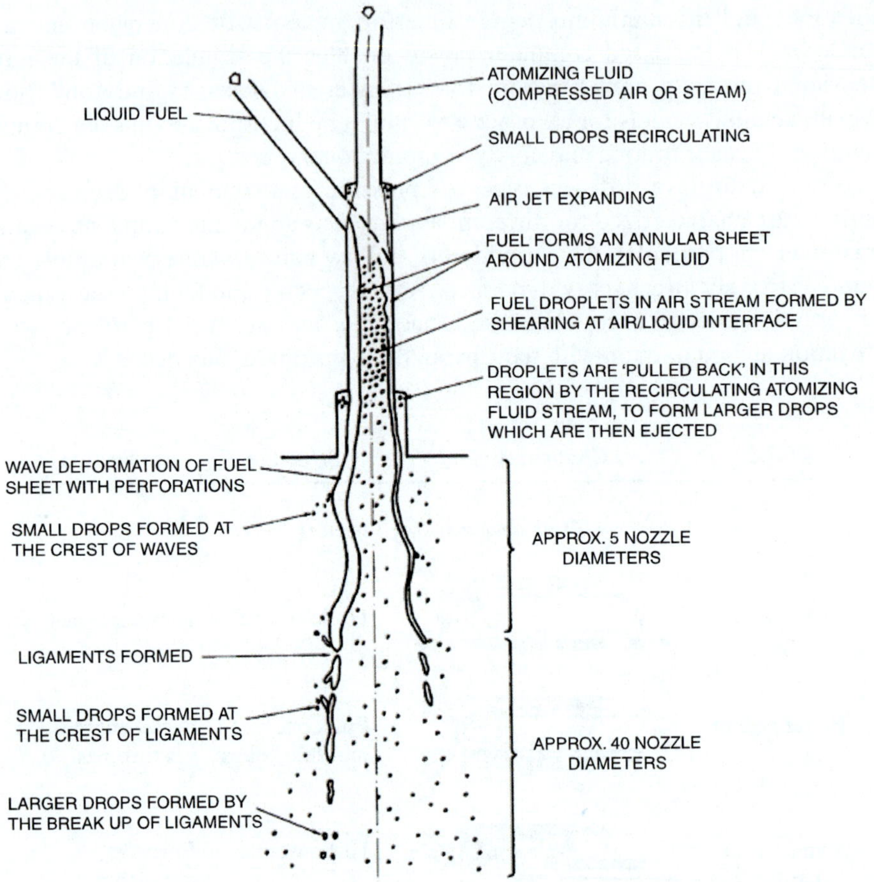

Fig. 1.3 Structure of a hollow cone spray (Reprinted with permission from Ref. [7]; © 1974, Maney Publishing)

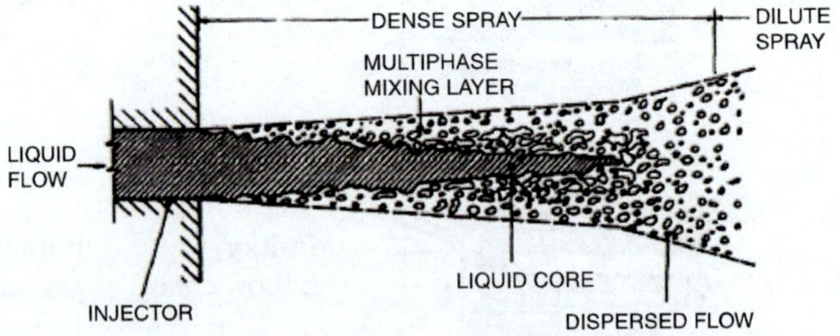

Fig. 1.4 Sketch of a near-injector region of a pressure-atomized spray in the atomization breakup regime (Reprinted with permission from Ref. [8]; © 1995, Elsevier)

Twin fluid and pressure atomizers produce quite different spray structures as can be seen in Figs. 1.3 and 1.4. The formation of fuel sheets and ligaments at the spray boundary close to the nozzle at about five nozzle diameters downstream are a challenge for numerical models whereas spray treatment of pressure atomizers as sketched in Fig. 1.4 is easier.

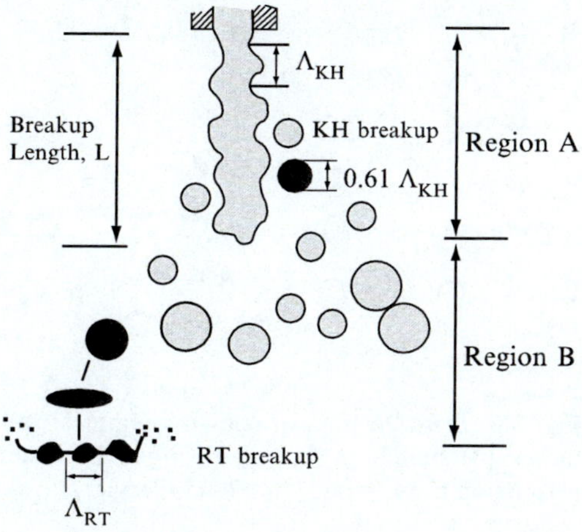

Fig. 1.5 Kelvin-Helmholtz (KH) and Rayleigh-Taylor (RT) instabilities

Figure 1.5 shows two main regimes during droplet breakup for full cone sprays. In region A, droplets detach from the intact liquid core, this process is called primary breakup. It mainly occurs through Kelvin-Helmholtz (KH) instability at the droplet surface, and it results from the relative velocity between the gas and the liquid phase. Once the droplet is detached from the intact liquid core, secondary breakup follows through Rayleigh-Taylor (RT) instabilities resulting from a dense, heavy fluid being accelerated by the light fluid.

Reitz's wave breakup model [2] is widely used to account for droplet breakup processes. This method considers the shearing-off of droplets due to the growth of KH instabilities on the droplet surface. Wavelength and growth rate of the instabilities are used to determine breakup time and details of the new droplets. Su et al. [3] improved the wave breakup model taking into account both the KH and RT instabilities.

Chryssakis and Assanis [4] give a unified fuel spray breakup model for internal engine combustion, where the breakup regimes are characterized through Weber and Ohnesorge numbers. Ko and Ryou [5] developed a new model for droplet collision-induced breakup accompanied by the formation of satellite droplets. Important regimes such as bounce, coalescence, and stretching as well as reflexive separation are identified and included. Different collision regimes are depicted in the right part of Fig. 1.6. Boundaries between these regimes are characterized through the Weber number, We, and the impact parameter, $b = 2B/(d_1 + d_2)$, where d_1 and d_2 are the droplet diameters of two colliding droplets, and B is the distance from the droplet center of one droplet to the relative velocity vector, placed on the center of the other droplet [5]. The model by Ko and Ryou [5] improves the widely used model for droplet-droplet collisions developed by O'Rourke [6]. The new model considers the collision-induced breakup process, and post-collision parameters of droplets are determined based on conservation equations between before and after collision.

Figure 1.7 shows numerical computations [9] where the influence of droplet evaporation on droplet collision is investigated. A comparison of row (c), displaying

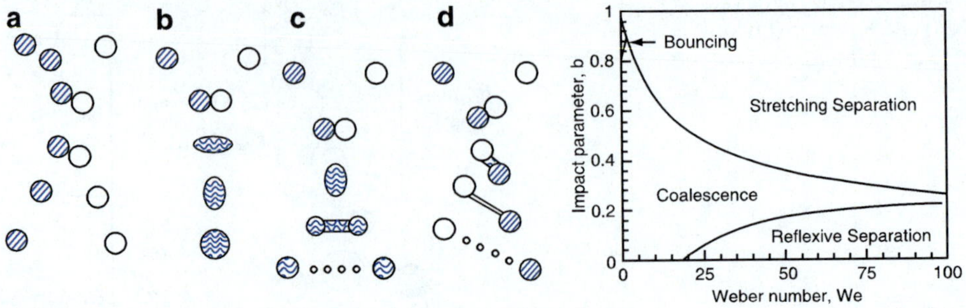

Fig. 1.6 *Left*: Diagram of collision regimes: (**a**) bouncing; (**b**) coalescence; (**c**) reflexive separation; (**d**) stretching separation. *Right*: Boundaries between collision regimes (Reprinted with permission from Ref. [5]; © 2005 Elsevier)

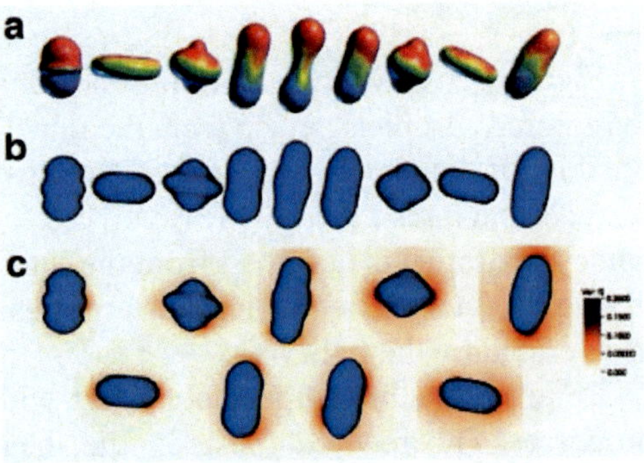

Fig. 1.7 Binary droplet collision for We = 20 without (**a**) and (**b**) and with (**c**) evaporation (Reprinted with permission from Ref. [9]; © 2008, Elsevier)

droplet collision with evaporation, with row (b) without evaporation, reveals that the influence of droplet evaporation on binary droplet collision is small for the case investigated.

Spray atomization and droplet breakup are the most challenging processes in spray modeling because of their dominating nature. Atomization characterizes the entire spray, and its effect on evaporation and combustion determines the entire technical process. Pressure atomizers are better understood than twin-fluid atomizers because of the increased complexity involved in the production of fuel sheets and ligaments characterizing twin-fluid atomization.

3 Droplet Evaporation and Combustion

Droplet evaporation and combustion are basic to both non-reactive and reactive spray flows. Single droplet evaporation and combustion have been studied since the mid of the last century, and basic findings such as the d^2 law [10, 11] which states that the droplet surface decreases linearly with time, ($\mathrm{d}d^2/\mathrm{d}t = -Kt$, where K is the evaporation constant), are still used in complex technical applications such as

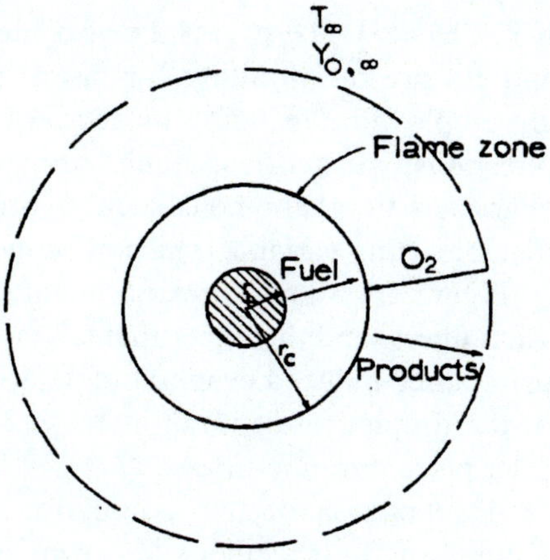

Fig. 1.8 Model of droplet burning in an oxidizing atmosphere (Reprinted with permission from Ref. [15]; © 1973, Elsevier)

diesel engine combustion [12]. The d^2 law includes assumptions [13–15] such as infinitely fast droplet heating and combustion in a thin flame sheet as well as quiescent, infinite oxidizing environment and constant liquid and gas properties, just to name the most important ones, cf. Fig. 1.8.

Some assumptions may be relaxed through simple approaches such as Sparrow's one third rule [16] to account for the temperature dependent physical properties in the immediate neighborhood of the droplet, others are more difficult to address, for instance the use of detailed chemistry [17–20].

The equations governing droplet heating, evaporation, and motion may be written for spherically symmetric droplets [14]

$$M_l c_{pl} \frac{\mathrm{d}T_l}{\mathrm{d}t} = \dot{m} \left[\frac{c_{pF}(T - T_l)}{B_T} - L_V(T_l) \right] \equiv Q_l \tag{1.1}$$

$$\dot{m} = 2\pi \rho_f D_f r_l \, \widetilde{\mathrm{Sh}} \ln(1 + \mathrm{B_M}) \tag{1.2}$$

$$\frac{\mathrm{d}\mathbf{v}}{\mathrm{d}t} = \frac{3}{8} \frac{\rho}{\rho_l} \frac{1}{r_l} (\mathbf{u} - \mathbf{v}) \mid \mathbf{u} - \mathbf{v} \mid C_D + \mathbf{g} \equiv \mathbf{F}_v, \tag{1.3}$$

where L_V is the temperature dependent latent heat of vaporization. B_M and B_T denote Spalding mass and heat transfer numbers, respectively.

The droplet evaporation rate, $\dot{m} = \frac{\mathrm{d}M_l}{\mathrm{d}t} = \frac{\mathrm{d}}{\mathrm{d}t}(\frac{4}{3}\pi\rho_l r_l^3)$ is computed for a droplet in a convective gas environment using the modified Sherwood number $\widetilde{\mathrm{Sh}}$ [14]. Equation 1.1 describes uniform droplet heating, i.e., the temperature is assumed to be spacially uniform. This assumption is valid for fuels with high volatility under low pressure and for small droplets with short droplet heating times compared to droplet life time. It can be relaxed through use of Eq. 1.4 for a spherically symmetric droplet [14]

$$\frac{\partial T_l}{\partial t} = \frac{\alpha_l}{r^2} \frac{\partial}{\partial r} \left(r^2 \frac{\partial T_l}{\partial r} \right), \tag{1.4}$$

where $\alpha_l = \lambda_l/(\rho_l c_{pl})$ is the heat exchange coefficient of the liquid. Equations 1.1 and 1.4 are rate limiting for fastest and slowest droplet heating, resp., and they are employed depending on the requirements of the application. In complex applications, most often, liquid properties are taken to be constant even though characteristic phenomena such as droplet expansion during heating are only predictable using variable transport properties [23].

Figure 1.9 shows an experimental setup of a suspended water droplet on two glass fibers, and the experimental and numerical results of this droplet evaporation using laser-induced evaporation is displayed in Fig. 1.10. Here, variable properties of the droplet result in an increase of droplet volume in particular for the longer laser pulse duration time.

A relaxation of the assumption of spherical symmetry requires the use of volume of fluid methods [24] which enable the consideration of deviation from one-dimensional treatment. Figure 1.11 shows a result of the volume of fluid

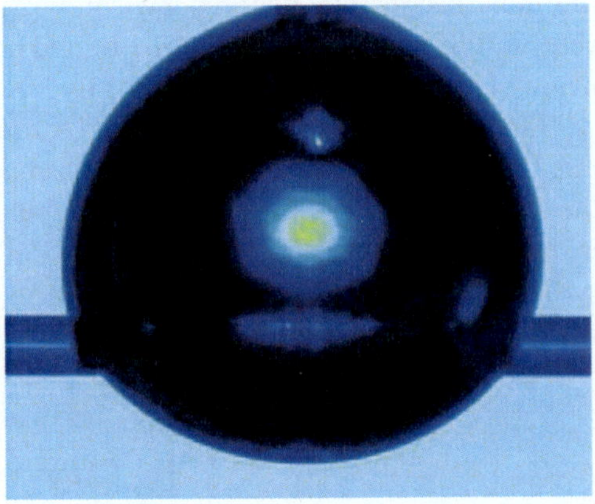

Fig. 1.9 Laser-induced droplet evaporation (Reprinted with permission from Ref. [21]; © 2004, Springer Science+Business Media)

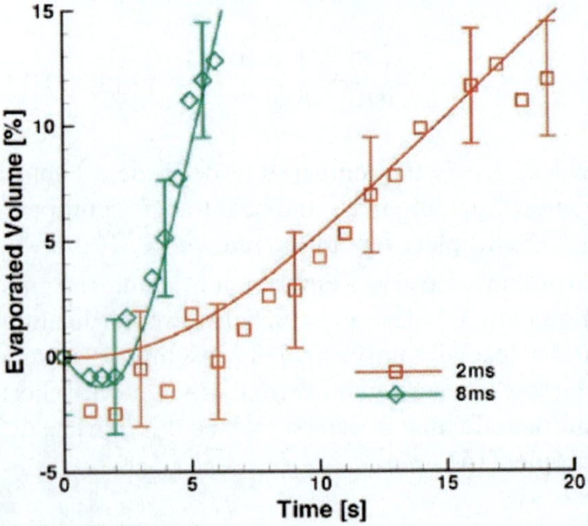

Fig. 1.10 Evaporated droplet volume versus time (Reprinted with permission from Ref. [22]; © 2007, Springer Science+Business Media)

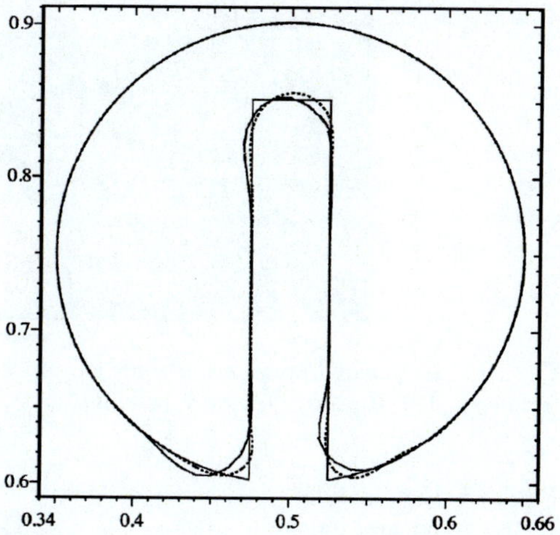

Fig. 1.11 Volume of fluid method for a distorted droplet (Reprinted with permission from Ref. [24]; © 2003, Elsevier)

method for a distorted droplet. The figure reflects the difficulties associated with sharp edges, and artificial smoothening is encountered in challenging cases as shown in the figure.

Droplets typically are released into a convective environment, and the effect of convection on both the droplet interior and the evaporation process itself has been studied extensively [14]. The effect of convection on the droplet interior is of particular interest in multi-component droplets where differences in physical properties of the liquid may lead to extraordinary phenomena such as micro-explosion [25]. In droplet and spray drying processes, differences in liquid properties may lead to different structures in the resulting particle [26].

In technical applications, droplets do not exist in an infinite oxidizing environment, but they are influenced by neighboring droplets. There is a number of papers that investigate two droplets side-by-side [27, 28] or in a tandem [18]. Scope of these papers is the investigation of the effects of hydrodynamic forces in presence of neighboring droplets where parameters such as droplet size and separation distance are studied. The flow field around the droplet changes with gas velocity and flow separation may occur downstream of the droplets. This may affect the ignition times as well as ignition positions in convective flow fields [18].

Figure 1.12 shows the ignition location downstream of a methanol droplet in air at elevated pressure where the ignition is shifted away from the axis of symmetry due to the flow separation in the droplet wake. Thus, droplet separation as seen in the figure has a pronounced effect on droplet ignition time and particularly on the position of droplet ignition. In situations where gas velocity is decreased [18], flow separation does not occur and ignition locates on the axis of symmetry.

A parameter study of single methanol droplet ignition displayed in Fig. 1.13 identifies minimum ignition times for droplet diameters of about 50 μm. Ignition times increase for smaller initial droplets due to gas-phase cooling during ignition and for larger initial droplet sizes due to prolonged droplet heating times.

More recent developments use molecular dynamics computations to study droplet evaporation in specific situations such as droplet evaporation under supercritical conditions [29] as shown in Fig. 1.14. Dark particles represent droplet molecules

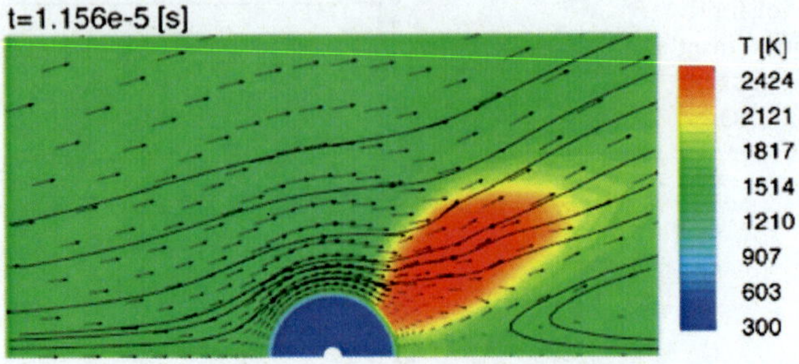

Fig. 1.12 Methanol droplet: zoom of the gas phase velocity at ignition (Reprinted with permission from Ref. [18]; © 2001, Taylor & Francis)

Fig. 1.13 Ignition time versus droplet size for different initial gas velocities (Reprinted with permission from Ref. [18]; © 2001, Taylor & Francis)

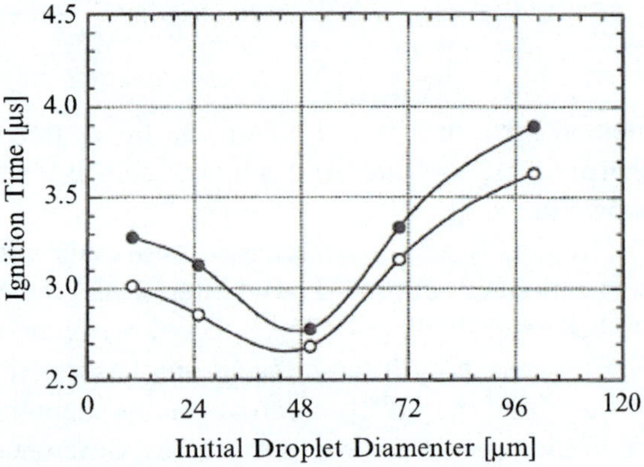

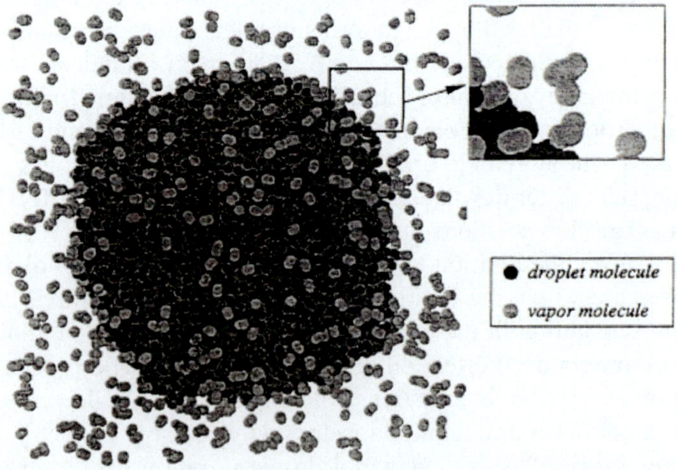

Fig. 1.14 Saturated liquid oxygen droplet at 100 K (Reprinted with permission from Ref. [29]; © 1998, Taylor & Francis)

and light particles are vapor molecules, respectively. The study concerns a sub-micron liquid oxygen droplet released into quiescent environments. In particular, conditions with vanishing surface tension are investigated.

Molecular dynamics is also used to study multi-particle collision [30], where molecular dynamics is combined with the mesoscale dynamics of the surrounding fluid. Figure 1.15 shows the flow field and the development of vortices behind a cylinder. The vortices seen in this approach may be qualitatively compared to the results of direct numerical simulation of the streamlines in and around a droplet in a convective flow field, see Fig. 1.16, where two droplets are placed side by

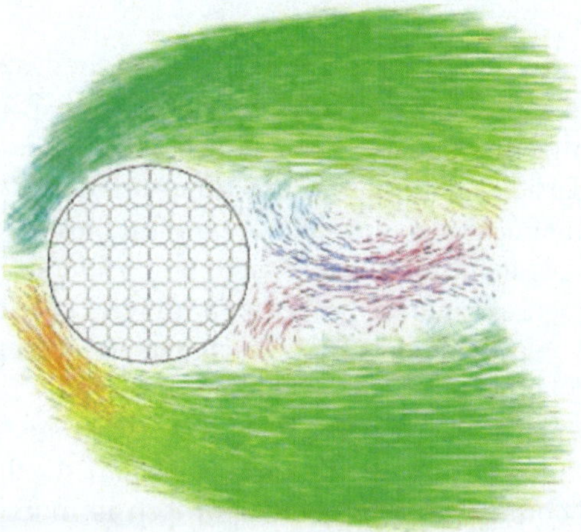

Fig. 1.15 Simulation of a flow field past a cylinder showing the development of vortices (Reprinted with permission from Ref. [30]; © 2004, Springer Science+Business Media)

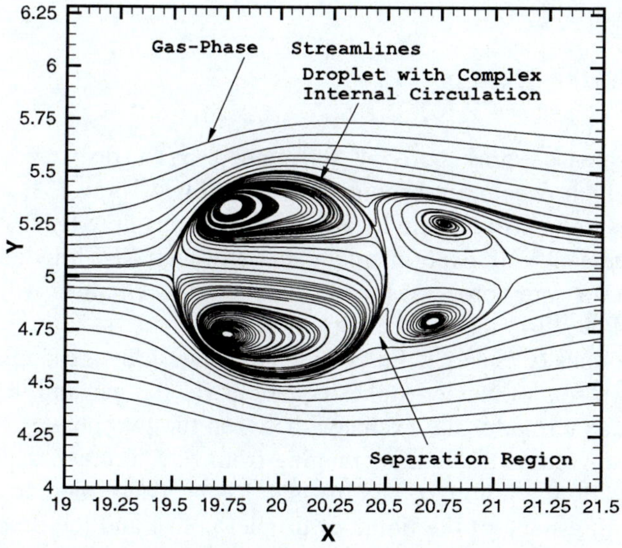

Fig. 1.16 Droplet in a convective flow field (Reprinted with permission from Ref. [28]; © 2004, Interscience Publishers)

side and the effect of the existence of neighboring droplets is studied [28]. Vortices and recirculation occur directly downstream the droplet, and they impose complex internal recirculation, which typically is neglected in simulations of technical applications due to their complex mathematical treatment which is associated with expensive computational cost.

More extended studies consider droplet streams or arrays, which are not addressed here.

4 Spray Combustion

In technical spray applications, spray evaporation and combustion typically occur in a turbulent flow where the cold spray is injected into an either cold or hot oxidizing environment. Major processes include the mathematical treatment of the spray particles and their interaction, turbulence and coupling with the liquid as well as chemical reactions and evaporation. Both turbulence and chemical reactions are issues themselves even without consideration of a liquid phase. The present section focuses on these processes with respect to their interaction with the liquid phase.

Spray models may be characterized depending on the density of the spray, defined as the ratio of volume covered by the liquid and the gas. For dense sprays, Euler-Euler models are appropriate whereas dilute sprays may be simulated using an Eulerian-Lagrangian approach where the dispersed spray is described in Lagrangian coordinates. The latter approach is more commonly used, but Eulerian-Eulerian models have become more significant of late.

4.1 Euler-Euler Models

Typically, the dense spray core is surrounded by a dispersed flow region, cf. Fig. 1.17, which eventually becomes a round dilute spray [31, 32]. The comparison of measurements with the locally-homogeneous flow (LHF) model shows qualitative agreement with experimental data. However, LHF models are not suitable to account for spray processes where detailed flow characteristics is required, and separated flow (SF) models are appropriate.

If the spray density is large, Euler-Euler methods [33] are applied which essentially use two sets of Navier-Stokes equations for the gas and the liquid phase containing source terms for the exchange between the two phases, and an indicator for the phase volume fraction, α, ranging from zero to unity, depending on the volume covered by the liquid, α, and the gas, $1 - \alpha$. These methods have the disadvantage that the shape of the liquid or droplet is lost and the description of the evaporation becomes more difficult compared to the case of a dilute spray.

More recent models not only treat the gas and the liquid phases as separate sets of equations, but the different components of the spray are accounted for through

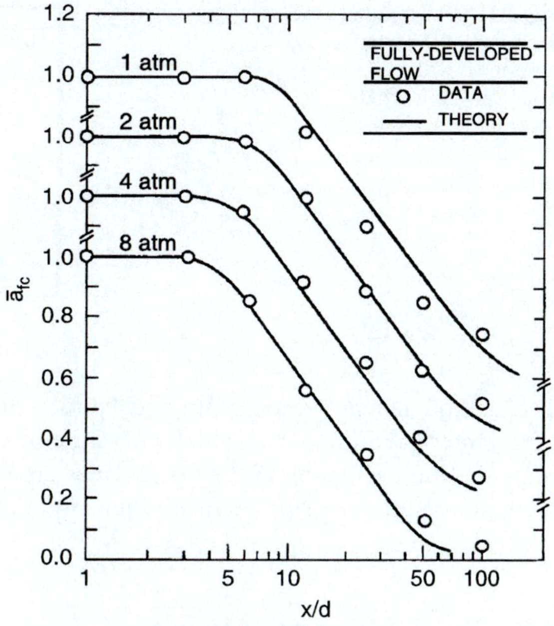

Fig. 1.17 Time-averaged liquid volume fractions along the axis of round, pressure-atomized water sprays in still air. Measurements and locally-homogeneous flow (LHF) predictions (Reprinted with permission from Ref. [31]; © 1992, American Institute of Aeronautics and Astronautics)

use of different sets of Navier-Stokes equations for the phases as well as the droplet size classes [34]. Thus, there is one phase for the gas and n phases for the liquid where n refers to the number of different droplet size groups [34] or to the different components of the liquid [33, 35]. These models require the definition of a liquid volume fraction, α_k, for the volume covered by phase k, over the total volume [34]:

$$\frac{\mathrm{d}(\alpha_k \rho_k)}{\mathrm{d}t} + \nabla(\alpha_k \rho_k \mathbf{v}_k) = \sum_{l=1, l \neq k}^{n} \Gamma_{kl}, \tag{1.5}$$

where the constraint $\sum \alpha_k = 1$ must be satisfied. Here, ρ_k denotes density, $\mathbf{v}_k$ the velocity vector of phase k, n the number of different liquid phases, and Γ_{kl} is the interfacial mass exchange term.

These equations must be solved together with continuity, momentum, and energy equations for the gas and the liquid phases as well as equations of state, interface conditions and equations for the interaction between the different size classes of the spray.

Another approach to simulate the dynamics of two-phase interfaces is the level set method [36]. Figure 1.18 illustrates the definition of the interface and the scalar field $G(\mathbf{x}, t)|_\Gamma$ [36] where Γ is the phase interface between two inviscid fluids, and the location and motion of the phase interface is captured by the level set equation

$$\frac{\partial G(\mathbf{x}, t)}{\partial t} + \mathbf{u}\,\nabla G(\mathbf{x}, t) = 0. \tag{1.6}$$

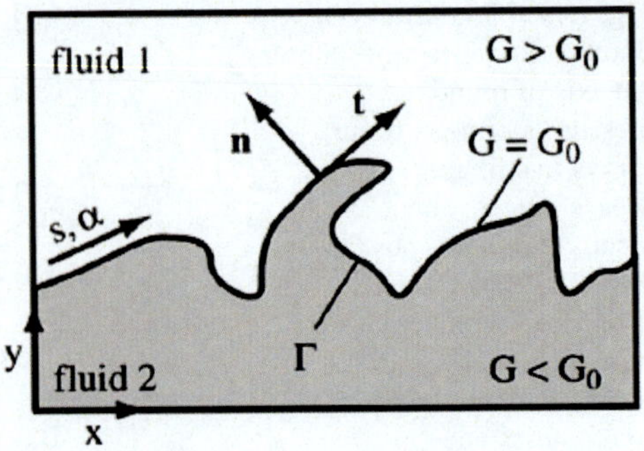

Fig. 1.18 Illustration of the definition of the interface (Reprinted with permission from Ref. [36]; © 2005, Elsevier)

The method includes specific numerical difficulties such as unintentional movement of the interface in regions of high curvature of G even if high order weighted essentially non-oscillating (WENO) schemes are employed, and different numerical methods to correct this problem are found in the literature [37, 38].

4.2 Euler-Lagrange Models

Models for dilute sprays typically use an Euler-Lagrange formulation where the dispersed phase is described by Lagrangian equations and the gas phase by Eulerian equations with source terms for mass, momentum, and energy, L_v, L_m, L_e, to account for the interaction with the spray. The instantaneous governing equations for mass, momentum in i direction, $i = 1, 2, 3$, for enthalpy, h_g, and for the mass fractions of chemical species, Y_i, $i = 1, \ldots, M$, may be written as

$$\frac{\partial \rho}{\partial t} = -\frac{\partial}{\partial x_k}(\rho u_k) + L_v \tag{1.7}$$

$$\frac{\partial(\rho u_i)}{\partial t} = -\frac{\partial(\rho u_k u_i)}{\partial x_k} - \frac{\partial \tau_{ki}}{\partial x_k} - \frac{\partial p}{\partial x_i} + \sum_{n=1}^{N} \rho_n \mathbf{b}_n + L_m \tag{1.8}$$

$$\frac{\partial(\rho h_g)}{\partial t} = -\frac{\partial}{\partial x_k}(\rho u_k h_g) - \frac{\partial \dot{\mathbf{q}}_{Ek}}{\partial x_k} + \sum_{n=1}^{N} \rho_n (\mathbf{u}_n - \mathbf{u})\mathbf{b}_n$$
$$+\frac{\partial p}{\partial t} + \frac{\partial(p u_k)}{\partial x_k} - \sum_{i=1}^{3}\sum_{j=1}^{3} \tau_{ij}\frac{\partial u_j}{\partial x_i} + L_e \tag{1.9}$$

$$\frac{\partial(\rho Y_i)}{\partial t} = -\frac{\partial(\rho Y_i u_{ik})}{\partial x_k} + M_i \dot{w}_i + \delta_{iF} L_v \tag{1.10}$$

where u_i are the gas velocity components, τ is the stress tensor, p pressure, ρ the gas density, **b** outer forces. Y_i and ω_i denote the mass fraction and molar chemical reaction rate of species i, respectively. The phase exchange terms L_v, L_m, and L_e may be written as [13]

$$L_v = \int\int 4\pi r_l^2 \rho_l R_l f \, dr_l \, d\mathbf{v} = -\int\int \dot{m} f \, dr_l \, d\mathbf{v} \quad (1.11)$$

$$L_m = -\int\int M_l \mathbf{F}_v f \, dr_l \, d\mathbf{v} - \int\int \dot{m} (\mathbf{u} - \mathbf{v}) f \, dr_l \, d\mathbf{v} \quad (1.12)$$

$$L_e = -\int\int \left(Q_l + \dot{m}[c_{pF}(T - T_{ls}) + L_V(T_{ls})]\right) f \, dr_l \, d\mathbf{v} - \int\int M_l \mathbf{F}_v \mathbf{v} f \, dr_l \, d\mathbf{v}. \quad (1.13)$$

In Eq. 1.12, $M_l = 4/3\pi r_l^3 \rho_l$ denotes the droplet mass. $\mathbf{F}_v$ is the drag force, cf. Eq. 1.3, and $f = f(r_l, \mathbf{x}, \mathbf{v}, t) \, dr_l \, d\mathbf{x} \, d\mathbf{v}$ denotes the distribution of the droplets in a spray, i.e. the probable number of droplets in the radius range dr_l around r_l within $d\mathbf{x}$ around $\mathbf{x}$ with velocity $d\mathbf{v}$ at $\mathbf{v}$ at time t.

Equations 1.1–1.3 or Eqs. 1.2–1.4 for droplet heating, vaporization, and motion discussed in Sect. 3 enter Eqs. 1.11–1.13 and therefore, the gas and liquid equations are strongly coupled. Moreover, the distribution function f must be specified. This may be achieved through several approaches. The droplet distribution function within a spray may be considered through discrete droplet models [14,39,40] which are quite well established. Here, the droplets are grouped in several size classes, and the droplet number density is obtained, cf. Fig. 1.19. This procedure is appropriate to account for droplet size distributions from experiment [41,42], since experimental methods such as particle image velocimetry (PIV) or phase Doppler particle analyzer (PDPA) result in discrete droplet size distributions.

The shortcoming of this approach is the fact that it is unknown how many size classes are needed to sufficiently represent the entire spray. On the one hand, a certain number is needed to describe the polydispersity of sprays, on the other hand the computational time increases with number of droplet size class. Currently, there is no mathematically profound criterion to determine the optimum number of droplet size groups in discrete droplet modeling. The Lagrangian part of numerical codes tremendously contributes to the computational time of a solver. Moreover, not only the number of size classes may be a problem but also the choice of width of the different size groups. Typically, equal-sized groups are considered.

Besides the discrete droplet model, the sectional approach is used where the development of Eulerian multi-fluid models for polydisperse evaporating sprays provides a promising tool. Figures 1.20 and 1.21 demonstrate the principal differences between the discrete droplet model (or method of classes) and the sectional approach. They are a compromise between precision and computational cost, and

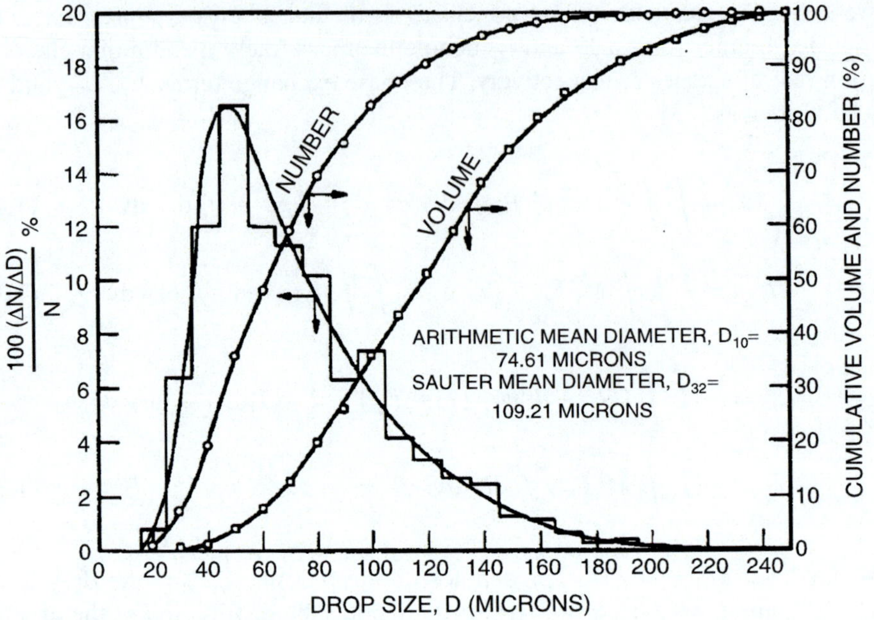

Fig. 1.19 Droplet distribution of a water spray in an air flow (Reprinted with permission from Ref. [1]; © 1977, Elsevier)

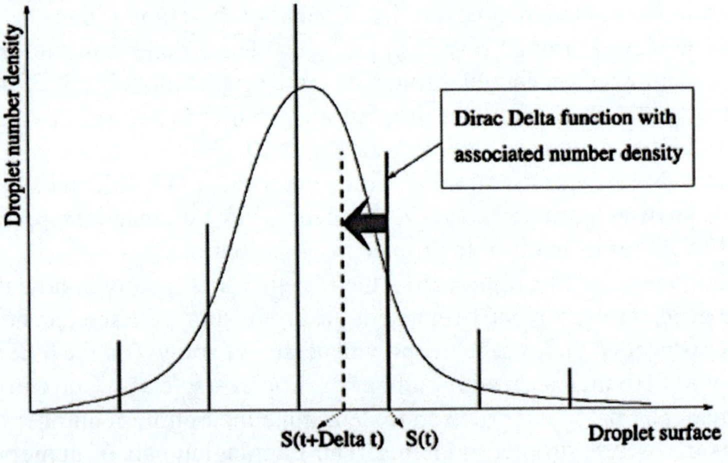

Fig. 1.20 Discrete droplet model or method of classes (Reprinted with permission from Ref. [43]; © 2001, Taylor & Francis)

competitive with stochastic Lagrangian methods. They provide a Eulerian description of the dispersed phase compatible with the gas phase description, thus allowing implicitation, vectorization or parallelization at relatively low computational cost. This is all the more true since the number of sections necessary in order to precisely

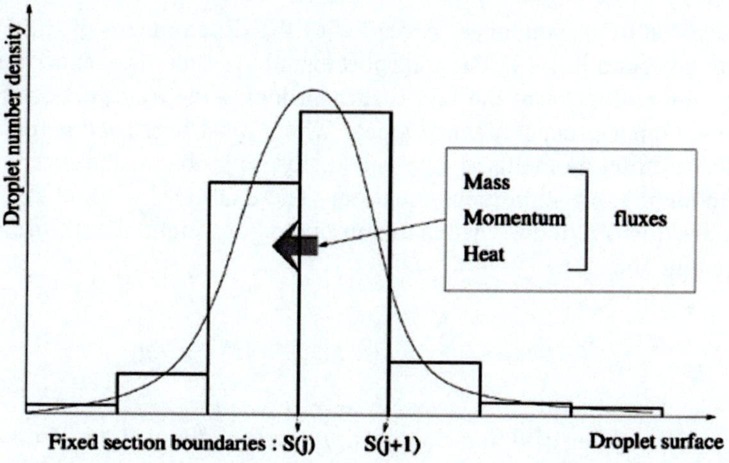

Fig. 1.21 Sectional approach (Reprinted with permission from Ref. [43]; © 2001, Taylor & Francis)

reproduce the vaporization process can be reduced to only a few [43]. It generates only one vaporization front for a given cloud, a feature that Lagrangian models do not have. Finally, since such models can be rigorously related to a kinetic formulation, they can be extended in order to take into account more complex phenomena such as coalescence or fragmentation [44], an uncommon feature of Eulerian models.

Another category of models to account for the spray processes within the Eulerian-Lagrangian description are PDF (probability density function) methods. Here, the droplet size distribution is assumed to be continuous and the spray equation yields [13, 45]

$$\frac{\partial f}{\partial t} = -\frac{\partial}{\partial r_l}(R_l f) - \frac{\partial (v_k f)}{\partial x_k} - \frac{\partial (F_k f)}{\partial v_k} + Q_f + \Gamma_f. \tag{1.14}$$

In Eq. 1.14, $R_l = \mathrm{d}r_l/\,\mathrm{d}t = -\dot{m}/(4\pi r_l^2 \rho_l)$ is the change of droplet radius, r_l, with time. Q_f denotes the velocity of the formation or destruction of particles through processes such as nucleation and breakup. The last term, Γ_f, describes changes due to droplet collisions. The last two terms must be considered in non-dilute sprays.

The numerical solution of this equation is particularly challenging when used in dense spray computations. A novel approach of Fox et al. [46, 48] developed for aerosols has a high potential to be applied in spray computations. The method belongs to the group of population balance theory and is called DQMOM (discrete quadrature method of moments).

DQMOM does not consider the probability density function (PDF) of the spray distribution itself, but it focuses on the first moments of the PDF. In this sense, the method is less precise than PDF methods since higher moments are neglected. However, both methods have shortcomings, and therefore, care must be taken to evaluate the advantages and disadvantages. A review of the development of the DQMOM method may be found in [46]. The transport equation of the moments of a particle size distribution suffers from the fact that in deducing the transport equations for the moments, higher-order moments appear which must be closed using relations including lower-order moments, the so-called closure problem. This causes insecurity in the modeling of the moment equations. The DQMOM method assumes that the particle size distribution is written as summation of weighted multi-dimensional Dirac–delta functions [46]

$$f(\xi;\mathbf{x},t) = \sum_{\alpha=1}^{N} \omega_\alpha(\mathbf{x},t)\delta[\xi - \langle\xi\rangle_\alpha(\mathbf{x},t)] \tag{1.15}$$

where N is the number of delta functions, $\omega_\alpha(\mathbf{x},t)$ is the weight of node α, and $\langle\xi\rangle_\alpha(\mathbf{x},t)$ is the property vector of node α. This presumed functional form may be thought as N distinct dispersed phases each of which is characterized by the weight factor and its property vector. This equation then is substituted into the population balance equation. With the definition of weights ω_α and abscissas, $\zeta_\alpha = \omega_\alpha\langle\xi\rangle_\alpha$, the DQMOM transport equations are written. The source terms of the transport equations of the weights, ω_α and the N weight abscissas, ζ_α, can be defined through a linear system involving the $2N$ integer moments, i.e., for $N = 3$, the moments $m_0, m_1, \ldots, m_5$ are obtained where these moments are defined as [46]

$$m_k = \int_{-\infty}^{\infty} \xi^k f(\xi)\,\mathrm{d}\xi = \sum_{\alpha-1}^{N} \omega_\alpha\langle\xi\rangle_\alpha^k. \tag{1.16}$$

The accuracy of the method increases with the increase of N, and a value of $N = 3$ is considered reasonable. It should be mentioned that sometimes the solution matrix becomes singular and the weights and abscissas do not represent an independent system of equations. They may also become negative, and numerical methods are discussed to prevent this non-physical behavior [47]. The advantage of the DQMOM method [48] is the ease to include processes of droplet interactions such as nucleation, aggregation, breakup, and growth [44]. A recent survey on population balance modeling of polydispersed particles in reactive flows is given by Rigopoulos [49].

4.3 *Turbulence Modeling*

Typically, technical applications of sprays occur under turbulent flow conditions in order to improve the mixing and to reduce the length of the spray core. Turbulence causes statistical fluctuations of all variables determining the flow field, and they need special numerical treatment.

The easiest and most precise but most computer time consuming method is direct numerical simulation (DNS) of turbulent (reactive) flows [50–52]. In this category of models, all turbulent length and time scales are resolved directly, and no modeling

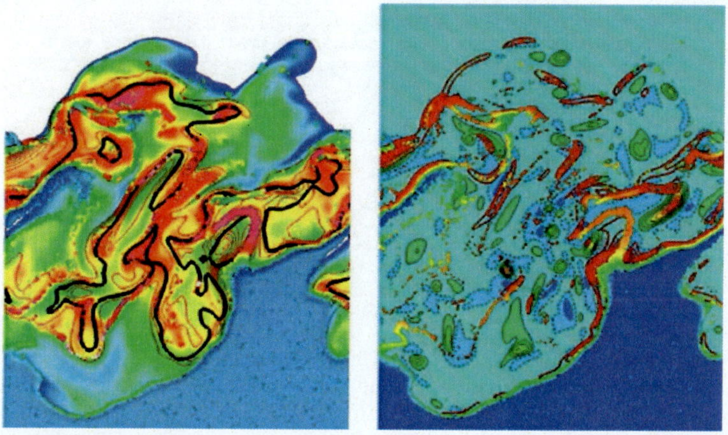

Fig. 1.22 DNS of diluted combustion of evaporating droplets (Reprinted with permission from Ref. [52]; © 2009, Elsevier)

of turbulent fluctuations is necessary. The disadvantage of the DNS is the high computational cost. Figure 1.22 displays results from DNS of diluted combustion with evaporating droplets [52] with focus on instantaneous interaction between evaporating droplets and reacting flow. The left part displays the reaction rate (red) and stoichiometric mixture fraction (black) as well as vapor mass fraction (flooded contours), whereas the right part shows the scalar dissipation rate (bold red lines) and the second invariant of the deformation tensor (flooded contours); droplets are superimposed. The strong interaction of all processes involved is evident. For complex technical applications, DNS is not suitable, and simplified model systems are considered [51, 52].

In large eddy simulation (LES) [53–55], large turbulence scales are resolved directly, and a filter function determines which smaller time scales are modeled using methods such as Reynolds averaged Navier Stokes (RANS) equations or probability density function (PDF) models, Reynolds stress (RSM) models, or conditional moment closure (CMC). Traditionally, the consideration of the liquid phase occurs within a Lagrangian framework whereas the continuous phase is described in an Eulerian system. A novel approach considers an Eulerian formulation for the spray which then is coupled with the LES solver for the gas phase [55]. Figure 1.23 shows the LES of an evaporating isopropyl alcohol spray using the dynamic Smagorinsky subgrid model [53] which seems to be the most common subgrid model currently used in connection with LES and evaporating sprays.

Models which refrain from directly resolving the large scales are RANS models, Reynolds stress models, PDF and CMC models. RANS models can be characterized by Reynolds averaging of the Navier Stokes equations where the instantaneous characteristic variables, Φ, first are replaced by the sum of their time or Favre averaged values, $\overline{\Phi}$ or $\tilde{\Phi}$, respectively, and their fluctuations, Φ' and Φ''. In turbulent reactive flows, Favre (density) averaged values need to be taken into account because of the high density changes associated with chemical reactions. After Reynolds averaging of the set of Navier Stokes equations, unclosed terms appear which have the form

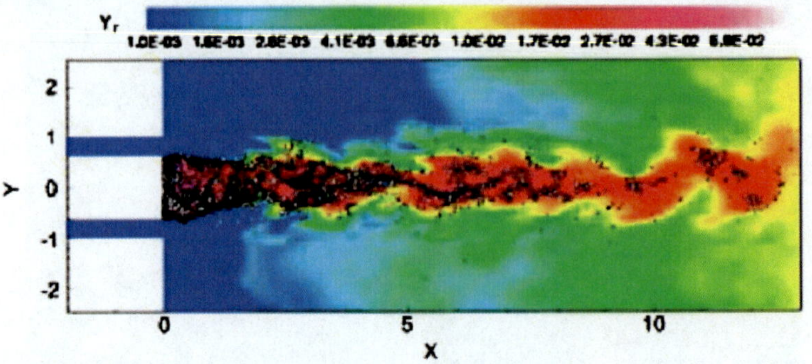

Fig. 1.23 LES of an evaporating isopropyl alcohol spray on unstructured grids using the dynamic Smagorinsky model (Reprinted with permission from Ref. [53]; © 2009, Elsevier)

$\overline{\rho u_k'' \Phi''}$, for Favre averaged values, the so called Reynolds stress terms. Here, ρ is the gas-phase density, u_k the gas-phase velocity component in k direction, and Φ denotes a characteristic variable such as total mass, enthalpy, gas velocity component, mixture fraction, or mass fraction of chemical species. The Reynolds stress terms then need to be closed by either a gradient flux approximation (this type includes the widely used $k - \epsilon$ turbulence model) or by derived transport equations for these terms. The latter approach leads to the Reynolds stress model where transport equations for the second moments are deduced and higher moments are closed using simplified assumptions leading to a closure of order two.

The major benefit of Reynolds stress models compared to gradient flux models is their capability of describing recirculation regimes as well as counter gradient diffusion. Models of the $k-\epsilon$ type additionally suffer from the assumption of isotropic turbulence. However, the gain of precision in the Reynolds stress models goes with additional computational cost for solving the additional transport equations of the second moments. Reynolds stress models are discussed below.

For turbulent spray flows, the correlation between the characteristic variables of the flow field and the evaporation rate needs special treatment in RANS models (for the $k-\epsilon$ model, see for instance Hollmann and Gutheil [40]). Considering turbulent mixing in a spray flow using a mixture fraction of the gas phase, ξ, defined such that it attains a value of unity for pure fuel and zero for pure oxidizer, the transport equation for the Favre mean, $\tilde{\xi}$, and the variance, $\widetilde{\xi''^2}$, of the mixture fraction, yield

$$\frac{\partial(\bar{\rho}\tilde{\xi})}{\partial t} + \frac{\partial(\bar{\rho}\tilde{u}_k\tilde{\xi})}{\partial x_k} - \frac{\partial}{\partial x_k}\left(\frac{\eta_{eff}}{\sigma_{\xi}}\frac{\partial\tilde{\xi}}{\partial x_k}\right) = \bar{L}_v \tag{1.17}$$

$$\begin{aligned}\frac{\partial\left(\bar{\rho}\widetilde{\xi''^2}\right)}{\partial t} &+ \frac{\partial}{\partial x_k}\left(\bar{\rho}\tilde{u}_k\widetilde{\xi''^2}\right) - \frac{\partial}{\partial x_k}\left(\frac{\eta_{eff}}{\sigma_{\widetilde{\xi''^2}}}\frac{\partial\widetilde{\xi''^2}}{\partial x_k}\right) \\ &= 2\left(\frac{\eta_{eff}}{\sigma_{\xi''2}}\right)\left(\frac{\partial\tilde{\xi}}{\partial x_k}\right)^2 - 2\bar{\rho}\frac{\tilde{\epsilon}}{\tilde{k}}\widetilde{\xi''^2} + \overline{2\xi'' L_v(1-\xi)} + \overline{\xi''^2 L_v},\end{aligned} \tag{1.18}$$

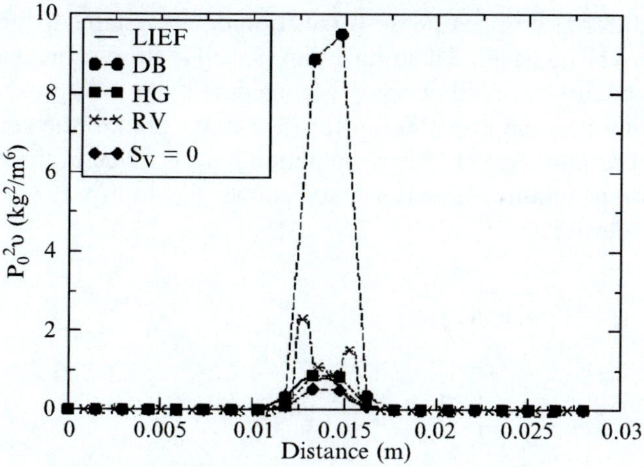

Fig. 1.24 Fuel vapor mass fluctuations at 2 cm and $t = 0.8$ ms (Reprinted from Ref. [56])

where the last two terms in Eq. 1.18 denote the interaction with the spray, L_v denotes the evaporated spray mass.

There are several approaches in the literature to close these terms. A survey and evaluation of current models is given by Subramanian et al. [56]. Figure 1.24 compares three models for the closure of the interaction of the turbulent mixing and the evaporation fluctuations. The model derived by Demoulin and Borghi [57] (DB) assumes that the vaporization is restricted to the droplet surrounding and the local mixture fraction is close to the saturation value. The model by Hollmann and Gutheil [40] (HG) assumes proportionality between the evaporated liquid mass and the mixture fraction as well as their turbulence intensities, this is a more global model which considers the influence of the evaporation not only in the near droplet surrounding. Réveillon and Vervisch [58] (RV) use an approximation of the conditional spray evaporation source term as monotonic function of mixture fraction. The numerical results from these models are compared with experimental data from laser induced exciplex fluorescence (LIEF) [56]. The authors conclude that the evaporation source term in the mixture fraction variance equation cannot be neglected [56]. Inside the core of the spray, its order of magnitude is close to the variance production and dissipation terms due to average mixture fraction gradient and scalar dissipation rate. In the experimental conditions used [56], the HG model performed better than the DB and RV models which both over-predict variance due to evaporation.

Modeling of the correlation terms of evaporation and gas velocity is required in Reynolds stress models [66]. Here the Favre-averaged Reynolds stress terms, $\overline{\rho u_i'' u_j''}$, are modeled through additional transport equations as discussed above.

The Reynolds stress model originally was successfully developed for turbulent recirculating gas flows [60], and it is extended [22] for use in spray flows. Moreover, Beishuizen [61] derived Reynolds stress equations with emphasis on the modeling of the pressure rate of strain tensor.

The major difference in gas-phase versus liquid/gas systems is the vaporization process that must be considered if atomization as well as droplet breakup and coalescence are neglected. Thus, a dilute spray is considered here. Moreover, the turbulent fluctuations caused by the flow field in reactive flows require the consideration of Favre-averaged equations [62]. The formulation here is given in three-dimensional coordinates, for generality. Equation 1.19 shows the transport equation for the Reynolds stress term [22]

$$
\begin{aligned}
&\frac{\partial}{\partial t}\left(\bar{\rho}\widetilde{u_i''u_j''}\right) + \frac{\partial}{\partial x_k}\left(\bar{\rho}\tilde{u}_k\widetilde{u_i''u_j''}\right) \\
&= \frac{\partial}{\partial x_k}\left(C_S\bar{\rho}\frac{\tilde{k}}{\tilde{\varepsilon}}\widetilde{u_k''u_l''}\frac{\partial\widetilde{u_i''u_j''}}{\partial x_l}\right) + P_{ij} - \frac{2}{3}\delta_{ij}\bar{\rho}\tilde{\varepsilon} \\
&\quad - C_1\bar{\rho}\tilde{\varepsilon}\left(\frac{\widetilde{u_i''u_j''}}{\tilde{k}} - \frac{2}{3}\delta_{ij}\right) - C_2\left(P_{ij} - \frac{1}{3}P_{ll}\right) \\
&\quad + C_3\left(\bar{\rho}g_i\overline{u_j''} + \bar{\rho}g_j\overline{u_i''} - \frac{2}{3}\delta_{ij}\bar{\rho}g_l\overline{u_l''}\right) - \frac{1}{4.3\bar{\rho}}\frac{\tilde{k}}{\tilde{\varepsilon}}\left(\frac{\partial\bar{p}}{\partial x_i}\widetilde{u_j''u_l''} + \frac{\partial\bar{p}}{\partial x_j}\widetilde{u_i''u_l''}\right)\frac{\partial\bar{\rho}}{\partial x_l} \\
&\quad - \overline{\left(u_i''u_j'' + \tilde{u}_i u_j'' + \tilde{u}_j u_i''\right)L_v} + \overline{u_j''L_{m,i}} + \overline{u_i''L_{m,j}}, \qquad (1.19)
\end{aligned}
$$

where the terms in the last line describe the interaction of the turbulent flow field with the vaporization source terms in the mass conservation equations, L_v, and the momentum equation, L_m, cf. Eqs. 1.11 and 1.12. All other terms are standard terms that also appear in gas-phase equations. These terms are closed following the procedure discussed in Ref. [63]. In the above equation, $P_{ij} = -\bar{\rho}\widetilde{u_j''u_k''}\frac{\partial\tilde{u}_i}{x_k} - \bar{\rho}\widetilde{u_i''u_k''}\frac{\partial\tilde{u}_j}{x_k}$ describes the production of Reynolds stresses through shear stress. The constants C_S, C_1, C_2, and C_3 are 0.22 [64], 3.0 [65], 0.33 [65], and 0.5 [59], respectively. δ_{ij} denotes the Dirac–delta function, and g_i is the gravitational force in i-direction.

The Reynolds-stress equation is solved together with the conservation equations for the momentum, the energy, and mass fractions of chemical species. Special care must be taken in discretizing the source terms of Eq. 1.19 in order to assure convergence [63].

Figure 1.25 displays a comparison [22, 63] of the Reynolds stress terms of a turbulent free methanol/air spray jet using the experiments of McDonell and Samuelson [41]. The agreement of the axial velocity fluctuations, $\sqrt{\widetilde{u''^2}}$ is quite good even though the Favre averaged mean value, $\tilde{u}$, is considerably lower in the experiment compared to the simulation near the axis. There are various reasons for this disagreement, and they are extensively discussed [40, 63, 67]. The major benefit of the Reynolds stress model is its capacity of predicting swirl as shown in Fig. 1.26, where eddies are predicted which cannot be computed using the $k - \epsilon$ model [22].

There is another category of turbulence models which avoid the mathematical modeling of the interaction of evaporation and other scalar shown in the previous

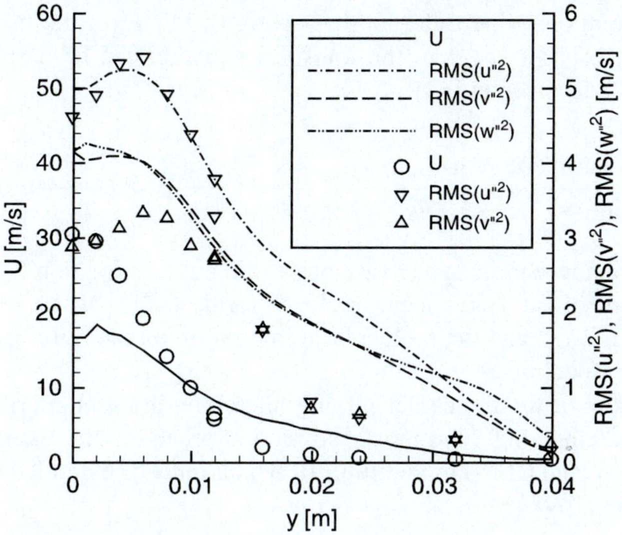

Fig. 1.25 Comparison of simulated (*lines*) [22, 63] and measured [41] (*symbols*) Reynolds stress terms at $x = 25$ mm (Reprinted with permission from Ref. [22]; © 2006, Springer Science+Business Media)

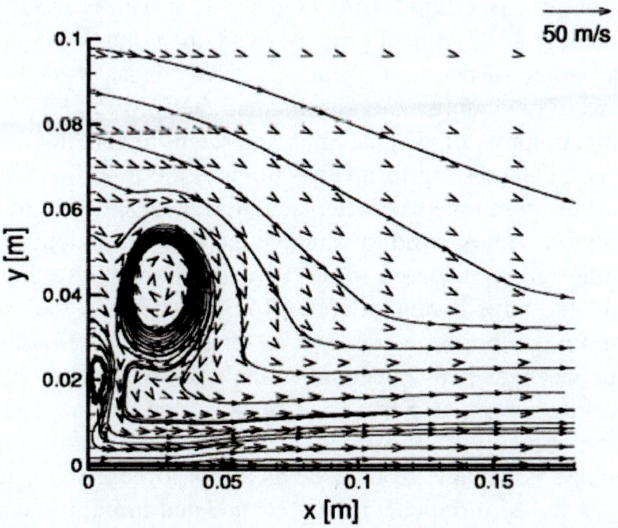

Fig. 1.26 Structure of a methanol/air spray flame: gas side of the spray flame, $a = 300$/s [22,63] (Reprinted with permission from Ref. [22]; © 2006, Springer Science+Business Media)

paragraph since the spray source term appears in closed term. The probability density function (PDF) models were introduced by Pope [68] for turbulent reactive gas flows. The advantages of the PDF model are exact treatment of convection, body forces, and mean pressure gradient, and chemical reactions, if present, cf. Sect. 4.4. More recently, the approach was extended to (non-reacting and reacting) spray flows since the spray source term appears in closed form, too [69].

For a one-point one-time Eulerian mass-weighted PDF for the mixture fraction $\tilde{f}(\zeta;\mathbf{x},t) = \rho(\zeta)\langle\delta(\xi-\zeta)/\langle\rho\rangle$, the transport equation can be derived following Pope's work [68] as given in [69]

$$\frac{\partial(\langle\rho\rangle\tilde{f})}{\partial t} + V_i\frac{\partial(\langle\rho\rangle\tilde{f})}{\partial x_i} + \frac{\partial(\langle L_v\rangle\tilde{f})}{\partial\zeta} = -\frac{\partial}{\partial\zeta}\left[\left\langle\frac{\partial}{\partial x_j}\left(\langle\rho\rangle D\frac{\partial\xi}{\partial x_j}\right)\middle|\zeta\right\rangle\tilde{f}\right]. \quad (1.20)$$

The variable ζ is the sample space variable of the mixture fraction ξ. The terms on the left hand side of Eq. 1.20 appear in closed form, this is particularly true for the spray source term, L_v, and the molecular mixing term on the right hand side of the equation needs modeling.

For the closure of the molecular mixing, there are different approaches derived from gas phase modeling. The most common approach is the interaction by exchange with the mean (IEM) model [68, 70] which must be extended to account for spray evaporation [69]

$$\frac{d\xi^*(t)}{dt} = -\frac{1}{2}C_\phi\frac{\tilde{\epsilon}}{\tilde{k}}(\xi^*(t)-\tilde{\xi}) + \frac{\bar{L}_v}{\langle\rho\rangle}, \quad (1.21)$$

where the last term is added to account for evaporation. Other models known are the Curl model [71] and the (modified) Curl model [72] as well as the Euclidean minimum spanning trees (EMST) model [73]. To the author's knowledge, the latter two models have not yet been used in spray flames but only in gas flows. In gas flows, the Curl model seems to give more exact results than the IEM model [72]. The EMST model is the most computer time consuming and seems to give the best results [73]. Therefore, it may be interesting to investigate both the modified Curl and EMST models to include evaporation effects for use in turbulent spray computations.

Joint PDF methods can be used to simulate the entire turbulent spray flow if all dependent variables are considered in the PDF. This procedure, however, is very costly because of the huge computational effort included. To the author's knowledge, the maximum number of dependents so far is two where particularly effects of correlations between the two dependent variables have been studied for both nonreacting and reacting sprays, cf. Sect. 4.4. Typically, flow characteristics are chosen that are particularly sensitive to the process as well as to the models involved.

The PDF method is suitable to be used as a tool to improve critical model assumptions. In gas flows, turbulent mixing is modeled through the assumption of presumed shapes of PDFs such as the β function,

$$P(\xi) = \frac{\Gamma(\alpha_1+\alpha_2)}{\Gamma(\alpha_1)\Gamma(\alpha_2)}\xi^{\alpha_1-1}(1-\xi)^{\alpha_2-1}, \quad (1.22)$$

where α_1 and α_2 are algebraic functions of $\tilde{\xi}$ and $\widetilde{\xi''^2}$ evaluated from transport equations (1.17) and (1.18), and Γ is the gamma function. A study by Miller and

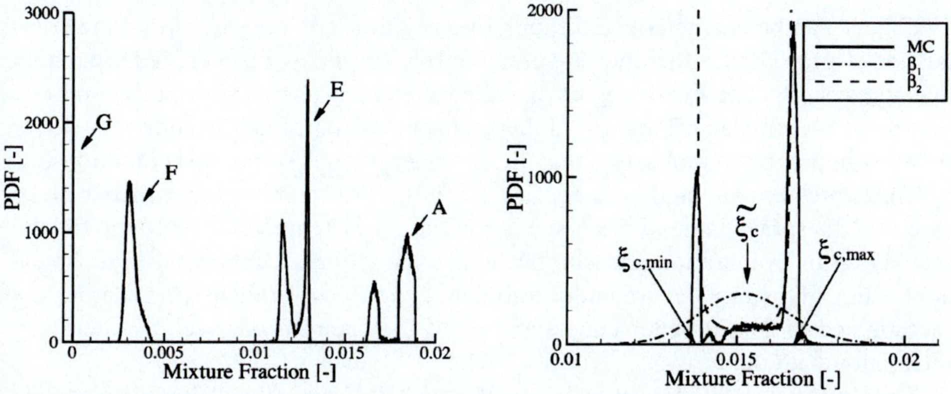

Fig. 1.27 PDFs of the mixture fraction of a methanol/air spray along the centerline (*left*) and comparison of the β PDF (β_1), modified β PDF (β_2) and Monte Carlo PDF (MC) (*right*) (Reprinted with permission from Ref. [69]; © 2006, Begell House, Inc.)

Bellan [74] shows that this presumed shape is not valid for spray flows. Studies by Ge and Gutheil [69, 75] show that an extension of the β function

$$P(\xi) = \frac{\Gamma(\alpha_1+\alpha_2)}{\Gamma(\alpha_1)\Gamma(\alpha_2)}(\xi_{\max}-\xi_{\min})^{1-\alpha_1-\alpha_2}(\xi-\xi_{\min})^{\alpha_1-1}(\xi_{\max}-\xi)^{\alpha_2-1}, \quad (1.23)$$

where two new parameters $\xi_{\min}$ and $\xi_{\max}$ were introduced, appears to overcome the shortcomings of the standard formulation. For the limits $\xi_{\min}=0$ and $\xi_{\max}=1$, the standard β function, Eq. 1.22, is retrieved as special case.

The new parameters, $\xi_{\min}$ and $\xi_{\max}$, shown in the right hand side of Fig. 1.27, are close to the values $\tilde{\xi} \pm \sqrt{\widetilde{\xi''^2}}$, however, this definition does not always lead to a good approximation, and further research is required.

PDF methods constitute a valuable tool for the evaluation of basic model assumptions as well as modeling of turbulent spray flows.

4.4 *Chemical Reactions in Spray Flows*

Chemical reactive flows have been the subject of many numerical studies, and the exponential temperature dependence of the Arrhenius rate expression requires special consideration in both laminar and turbulent flames. Fuels such as kerosene or diesel consist of many components, and the chemical reaction mechanisms are not well known. Detailed chemical reaction mechanisms consisting of elementary reactions are very long and chemical reaction rate constants again, are not well known. For combustion problems, the CHEMKIN code [76] was developed by Sandia National Laboratories in CA, USA. CHEMKIN is a data base for thermodynamic data as well as reaction rate constants for elementary kinetic reaction steps for a big variety of gas-phase fuels.

Flames can be characterized through the way they are introduced into the combustion chamber, i.e., either they are premixed or fuel and oxidizer are fed separately into the system. The first category is called premixed flames and the second one non-premixed, the latter having a higher impact in technical applications for safety reasons. In practice, combustion may occur under partially premixed conditions.

There are several model categories for gas-phase combustion which can be extended to spray flames where special attention is required to account for the interaction of evaporation and combustion. Even though combustion in technical applications typically occurs under turbulent conditions, laminar spray flames are investigated to gain a better understanding of the chemistry as well as chemistry interaction with the spray.

The counterflow configuration is a well-established setup which has been studied extensively for both gas and spray flames. A paper by Li [77] gives an excellent survey on this issue until the year 1997, but thereafter, several other groups have published in this field [78–84]. The chemical reaction mechanisms typically are taken from gas-phase models such as the GRI mechanism [85] or Warnatz's mechanism [86]. Laminar spray flame computations most often are performed on two-dimensional [79] or one-dimensional [78, 81–84] grids, with the latter ones resulting from transformation using similarity transformations [87]. Therefore, the use of detailed chemical reaction mechanisms is affordable. At the same time, these configurations have been used to reduce chemical reaction mechanisms by both steady-state assumptions of chemical species and assumptions of partial equilibrium of certain elementary reactions [88]. An overview of systematically reduced chemical reaction mechanisms is given by Peters and Rogg [89].

A more recent method of reducing detailed chemical reactions is the method of intrinsic low-dimensional manifolds (ILDM). The ILDM approach uses the fact that in typical reaction systems, a large number of chemical processes are so fast that they are not rate limiting and can be decoupled. Intrinsic low-dimensional manifolds in the state space are identified with the property that after a short relaxation time the thermochemical state of the system has relaxed onto these attracting low-dimensional manifolds [90]. This approach, however, to the author's knowledge, has not yet been introduced into the simulation of spray flames.

Flamelet models are extensively used in gas flame computations [91, 92], and this approach was extended to spray flames [42, 67, 69, 75]. Most publications include gas flamelets and ignore the effect of spray evaporation on the structure of the flamelet [40]. More recently, spray flamelets have been studied [78, 81–84] to account for flame interaction with the spray, and they are also used in turbulent spray flame computations [42, 67, 75]. The basic idea followed in flamelet models is that the combustion is determined by the mixing through the mixture fraction, ξ, and its scalar dissipation rate, χ, where the instantaneous scalar dissipation rate is defined as

$$\chi = 2D\left(\frac{\partial \xi}{\partial x_j}\right)^2. \tag{1.24}$$

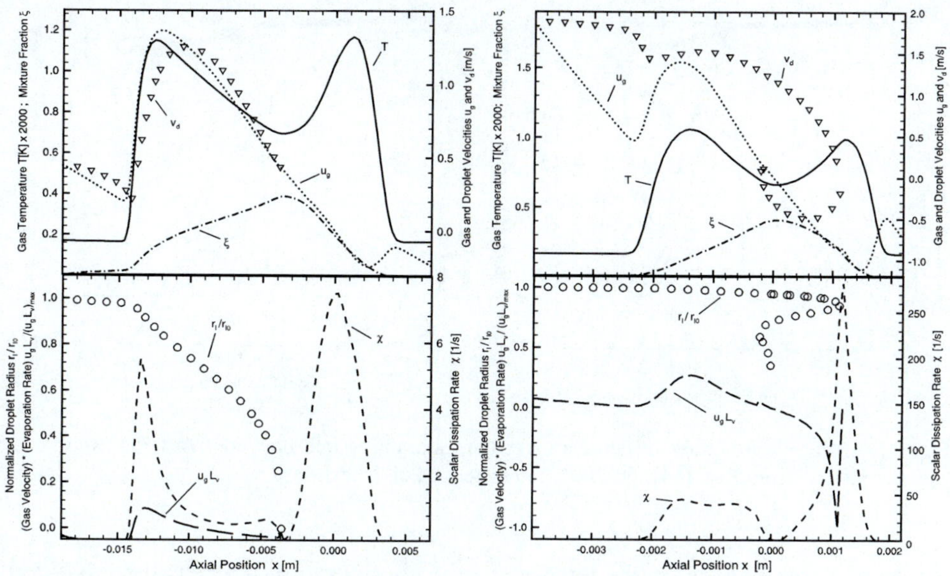

Fig. 1.28 Structure of laminar spray flames in the counterflow configuration for low (*left*) and high (*right*) strain rate (Reprinted with permission from Ref. [67]; © 1998, Taylor & Francis)

All other variables, Φ, such as temperature and mass fractions of chemical species, then are evaluated from ξ and χ through

$$\Phi = \Phi(\xi, \chi). \tag{1.25}$$

These dependencies are tabulated for use in turbulent flame computations.

However, laminar spray flames have considerably different structures and dependencies from gas flames. Figure 1.28 gives insight into characteristics of laminar spray flame structures where two different strain rates of 55/s (left) and 1,000/s (right) [67] are shown. A liquid monodisperse methanol spray enters from the left with carrier gas air, and it is directed against an air stream leading to a double flame where the left is a spray flame and the right hand side flame is a gas diffusion flame. Flame temperature decreases with increased strain and the droplets reverse and oscillate around the stagnation plane at axial position zero. Drag causes droplet velocity to strongly differ from gas velocity, and differences increase with higher strain. The mixture fraction, ξ, shows a non-monotonic behavior: in contrast to pure gas flames, two values of mixture fraction at stoichiometric condition, ξ_{st} are attained which correspond to the two flames characterizing the spray flames investigated here. Also, scalar dissipation rate, χ, attains two maxima which is different from one maximum found in gas combustion. The second maximum found on the spray side corresponds to evaporation, where L_v denotes the evaporated liquid mass.

Moreover, multiple solutions [84] of laminar spray flames in the counterflow configuration were found to exist at low strain rate. In both solutions shown in Fig. 1.29,

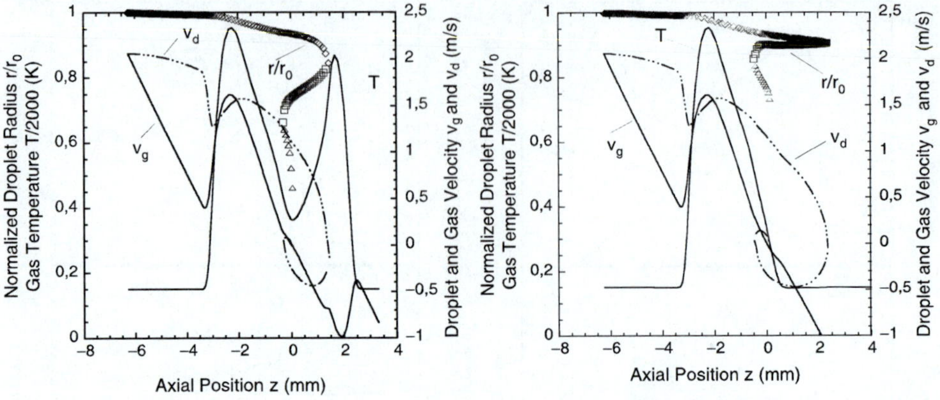

Fig. 1.29 Multiple structures of laminar spray flames in the counterflow configuration (Reprinted with permission from Ref. [84]; © 2005, Interscience Publisher)

the spray flame resides on the left side of the configuration whereas the gas-sided reaction zone does not exist in the right part of Fig. 1.29. Existence of multiple solutions was postulated earlier [78], but they had not been identified. The non-existence of the gas flame corresponds to the "cold solution" in pure gas combustion which essentially reflects mixing of the reactants in pure gas situations. Figure 1.29 also demonstrates the effect of the gas-sided flame on the droplets which are oscillating around the stagnation plane. Multiple solutions obtained for low strain rates up to 500/s do not persist at higher strain because the reaction zones merge due to reduction of flame thickness. At the same time, it was argued [84] that these multiple solutions and the similarity of the gas-sided reaction zone to pure gas flame structures facilitate flamelet computations in turbulent spray combustion modeling.

The effect of turbulent fluctuations on laminar flamelets is accounted for through use of probability density functions. In the case of gas flames, the dependent variables of the PDFs are mixture fraction, ξ, and its scalar dissipation rate, χ, cf. Eq. 1.25. The turbulent (Favre averaged) value of a variable $\tilde{\Phi}$ in the turbulent flow field then is evaluated through integration over the laminar value weighted by the Favre averaged PDF $\tilde{P}(\xi, \chi)$. For turbulent spray flames, the approach must be extended to account for the dependence of the laminar spray flamelet on the initial droplet size, r_0, and initial spray velocity, v_0, and the equivalence ratio, E_r, i.e.,

$$\Phi = \Phi(\xi, \chi, r_0, v_0, E_r). \tag{1.26}$$

The Favre averaged value of any variable $\tilde{\Phi}$ in the turbulent flow field then depends on five variables, and it yields [67]

$$\tilde{\Phi}(\xi, \chi_{\mathrm{st}}, r_0, v_0, E_r) = \int_0^\infty \int_{-\infty}^\infty \int_0^\infty \int_0^\infty \int_0^1 \Phi(\xi, \chi_{\mathrm{st}}, r_0, v_0, E_r) \times \tilde{P}(\xi, \chi_{\mathrm{st}}, r_0, v_0, E_r)\, \mathrm{d}\xi\, \mathrm{d}\chi_{\mathrm{st}}\, \mathrm{d}r_0\, \mathrm{d}v_0\, \mathrm{d}E_r, \tag{1.27}$$

where the scalar dissipation rate, χ, is replaced by its stoichiometric value, χ_{st}, which is the standard procedure in turbulent flamelet computations [67, 91].

The joint PDF depending on five variables is factorized, and simplified presumed shapes of unimodal PDFs are used to describe these marginal PDFs. For the mixture fraction, $\tilde{P}(\xi)$, a β function or modified β function is used, cf. Eqs. 1.22 and 1.23 [67, 75]. For the scalar dissipation rate at stoichiometry, χ_{st}, a logarithmic normal distribution with standard deviation two is used [93]. For the initial droplet size, two Dirac–delta functions were used, and the PDFs of both the initial droplet velocity and equivalence ratio were taken to be univariate Dirac–delta functions with initial droplet velocity and global equivalence ratio [67]. These assumptions as well as the assumption of statistical independence may not be very good choices even though it has been explored that the choice of one or two Dirac–delta functions for the droplet size does not show significant influence on the results of numerical simulations [63, 67].

Figure 1.30 shows a comparison of the flame structure using a gas flamelet library (left) and a spray flamelet library (right), and major differences are obvious. Artificial lift-off is predicted by the gas flame library and corrugation at the flame boundary cannot be predicted neglecting droplets. The latter is a consequence of large droplets crossing the flame front.

Another method to include chemical reaction mechanisms are flamelet generated manifolds [94]. Here, the chemical reactions are parametrized through manifolds

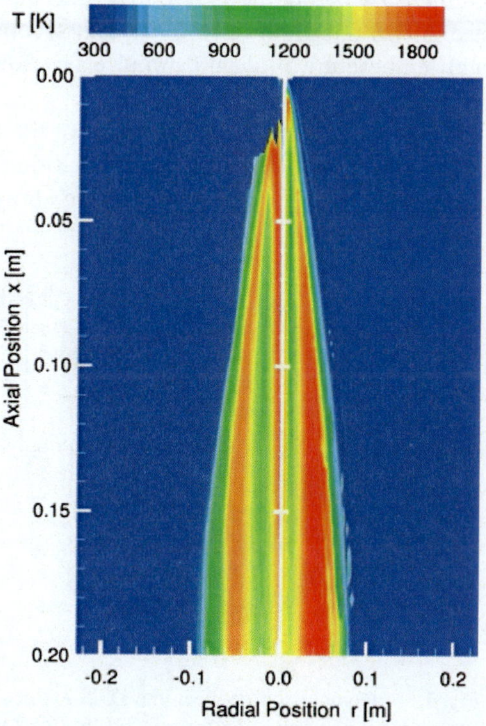

Fig. 1.30 Structure of a methanol/air flame: gas flamelet (*left*) versus spray flamelet (*right*) (Reprinted with permission from Ref. [22]; © 2006, Springer Science+Business Media)

that are obtained from laminar flamelet computations [95] as described above. This is essentially a combination of ILDM and flamelet modeling where the chemistry is taken from the flamelet computations and the tabulation method from ILDM. This method of considering chemical reaction mechanisms in spray flames is not yet a standard procedure either, and more research is warranted.

If spray flames are turbulent, the chemical rate expressions which appear in Arrhenius form, include exponential temperature dependence, and the turbulent fluctuations of temperature may greatly influence the chemical reaction rates. Direct closure methods use assumed shapes of probability density functions to account for the turbulent fluctuations in the turbulent flow field and their influence on the instantaneous chemical reaction rate, $\dot{\omega}_i$, for species i, compare Eq. 1.10

$$\overline{\dot{w}_i} = \int_0^\infty \int_0^\infty \int_0^1 \ldots \int_0^1 \dot{w}_i(Y_1, \ldots Y_M, T, \rho) P(Y_1, \ldots Y_M, T, \rho)\, \mathrm{d}Y_1 \ldots \mathrm{d}Y_M\, \mathrm{d}T\, \mathrm{d}\rho, \tag{1.28}$$

where $P(Y_1, \ldots Y_M, T, \rho)$ is a multivariate joint probability density function. In practice, this PDF is not suitable because of the numerical difficulties involved. However, if simplified chemistry as discussed above is used, the PDF depends only on a few variables, and after factorization of the function (assuming statistical independence of the dependent variables), presumed shapes of the PDF such as the β function or modified β function, may be employed.

Recent studies by Miller and Bellan [74], Selle and Bellan [96] and Ge and Gutheil [69, 75] demonstrate that first, the presumed β function is not suitable for use in two-phase flows [69, 74], and second, the variables such as evaporation rate and temperature [96] and mixture fraction and enthalpy [75] are not statistically independent.

Figure 1.31 shows a comparison of the DNS-extracted PDF and the model PDF at the subgrid scale of a LES computation for the gas temperature, T'' and the vapor mass fraction, Y_v'' for a non-reacting spray taken from Selle and Bellan [96]. For

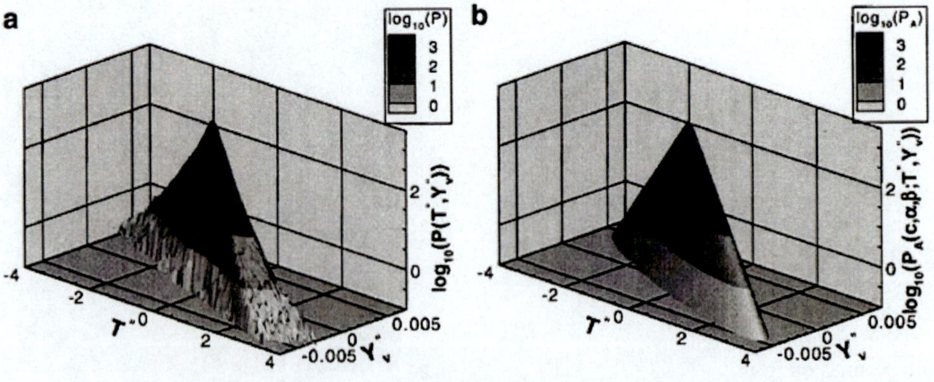

Fig. 1.31 Comparison between the DNS-extracted PDF and the model PDF at the subgrid scale (Reprinted with permission from Ref. [96]; © 2007, Elsevier)

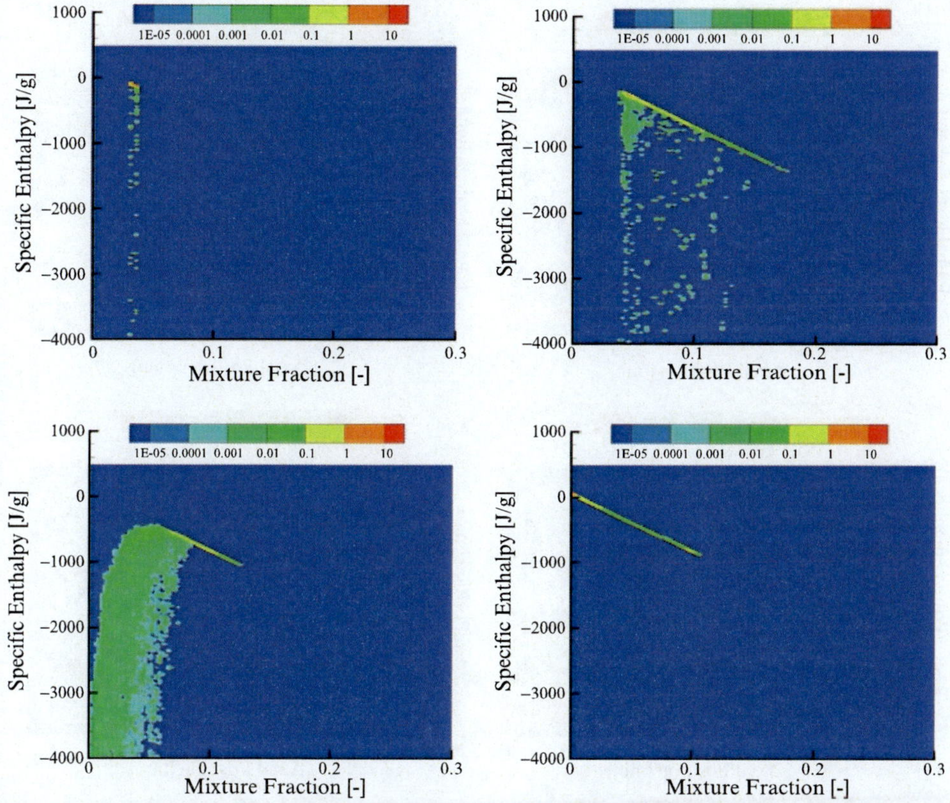

Fig. 1.32 Contour plot of joint enthalpy-mixture fraction PDF at various positions (Reprinted with permission from Ref. [75]; © 2008, Elsevier)

pure gas flows, the variables are uncorrelated whereas for the two-phase flow, a clear correlation is displayed [96]. The right hand side of Fig. 1.31 is obtained from a primitive of a modified Bessel function of the second kind [96].

Another way of checking statistical independence is use of the joint PDF transport equation for two variables of interest. The joint PDF for the mixture fraction and specific enthalpy [42, 75] is used to evaluate the correlation coefficients at different positions in reacting turbulent methanol [75] and ethanol [42] sprays in air.

Figure 1.32 displays contour plots of the mixture fraction and specific enthalpy in the region where evaporation starts but no mixing occurs (top left), here the variables are statistically independent. In areas where both evaporation and combustion occur (top right and bottom left), the variables are strongly correlated, and after evaporation is finished, a linear dependence is obtained, see Fig. 1.32 (bottom right), which is in agreement with the findings of Selle and Bellan [96] for reacting turbulent sprays.

The spray flamelet model in connection with the joint PDF for mixture fraction and enthalpy enables prediction of chemical species displayed in Fig. 1.33.

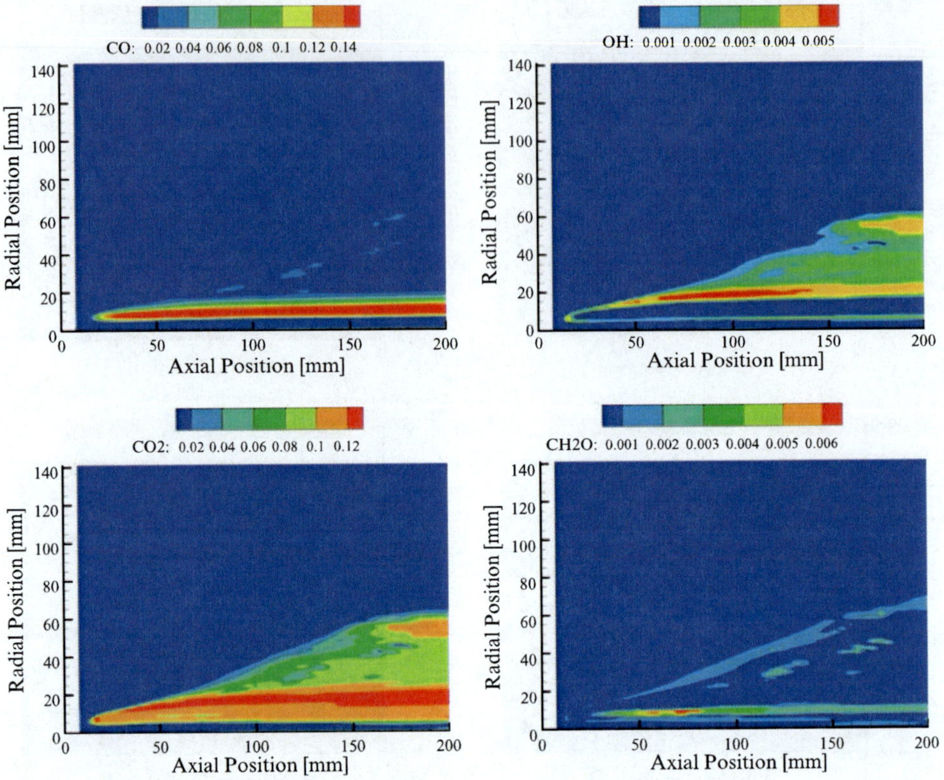

Fig. 1.33 Contour plot of CO, OH, CO_2, and CH_2O in a turbulent methanol/air spray flame (Reprinted with permission from Ref. [75]; © 2008, Elsevier)

In hot regions with high CO and OH values, the rate-limiting reaction $CO + OH \rightleftharpoons CO_2 + H$ with its high activation energy determines the production of CO_2. Moreover, formaldehyde can be predicted which is unstable for high-temperature regions as can be seen in Fig. 1.33.

A different type of models to predict turbulent mixing and combustion is conditional moment closure (CMC) for turbulent combustion. This approach was originally introduced by Klimenko and Bilger [97] for gas combustion: Here the conditional averages of higher moments of quantities such as species mass fractions and enthalpy, conditional on the mixture fraction in diffusion flames, or conditional on the reaction progress variable in premixed combustion are considered.
Conditioning on the mixture fraction ξ yields [98]

$$\langle Y_i \rangle = \int_{-\infty}^{\infty} \xi(\eta) \langle Y_i | \eta \rangle d\eta, \tag{1.29}$$

where η is the sample space variable for ξ.

Mortensen and Bilger [98] also give an outline for use of CMC for spray flames, where double-conditioning should be used [97], $Q \equiv \langle Y | \xi = \eta, \phi = f \rangle$. The variations of the mixture fraction, ξ, are related to turbulent macro-transport, and the variations of ϕ are related to inter-droplet micro-processes. The latter process is considered more relevant, and equations are derived excluding macro-fluctuations. Equations are given for homogenous turbulence with ξ = const. and $Q \equiv \langle Y | \phi = f \rangle$. Mortensen and Bilger [98] derive a formulation of CMC for spray combustion in connection with a level-set function tracking the interface between gas and liquid. Moreover, an indicator function is defined for identification of gas or liquid. Identical equations for the two phases are derived, and it is shown that additional terms arise to account for dynamic boundary conditions between the phases.

The treatment of chemical reactions within the framework of CMC is discussed by Merci et al. [99]. In first-order CMC, it is assumed that the conditional chemical reaction rate is $\langle \dot{\omega}_i | \xi \rangle = f(\langle Y_1 | \xi \rangle, \ldots, \langle Y_N | \xi \rangle), \langle T | \xi \rangle)$ neglecting turbulent fluctuations. This can be relaxed through use of second-order conditioning on both mixture fraction and its variance or through double conditioning, for instance on the mixture fraction and enthalpy. So far, CMC has been applied to RANS simulations [100], and studies of combined LES/CMC have not been reported.

It can be seen that models for pure gas flames may not be transferred to spray flames because of the impact of the liquid phase on the flame structure. Droplet and spray distribution greatly influence the entire flow field and combustion process, and special care must be taken if gas-phase models are extended to sprays.

5 Summary and Conclusions

There is a number of open issues in droplet and spray modeling and simulation. They include droplet interaction and integration of these models into simulations of technical spray flames. Moreover, the statistical distribution of droplets within the spray is not sufficiently described by current methods, and further research is warranted.

Research on the effect of turbulence on spray flames is in progress, also due to the fact that nowadays computers become increasingly fast, and both DNS and LES become more attractive. LES modeling suffers from a lack of suitable subgrid models where special focus on statistical correlations on characteristic variables such as mixing and evaporation is required.

Moreover, a number of open issues such as consideration of advanced chemical reaction schemes for use in spray flows must be addressed. They include chemical reaction mechanisms for alternative fuels such as biological fuels, which was not addressed here.

References

1. Faeth, G.M.: Current status of droplet and liquid combustion. Prog. Energy Combust. Sci. **3**, 191–224 (1977)
2. Reitz, R.D.: Modeling atomization processes in high-pressure vaporizing sprays. Atomisation Spray Technol. **3**, 309 (1987)

3. Su, T.F., Patterson, M.A., Reitz, R.D., Farrell, P.V.: Experimental and numerical studies of high-pressure multiple injection sprays. SAE Technical Paper Series, Paper 960861, Society of Automotive Engineers, Inc., Warrendale, PA (1996)
4. Chryssakis, C., Assanis, D.N.: A unified fuel spray breakup model for internal combustion engine applications. Atomiz. Sprays **18**, 1–52 (2008)
5. Ko, G.H., Ryou, H.S.: Droplet collision processes in an inter-spray impingement system. Aerosol Sci. **36**, 1300–1321 (2005)
6. O'Rourke, P.J.: Convective droplet effects on vaporizing liquid sprays. Ph.D. thesis, Mechanical and Aerospace Engineering, Princeton University (1981)
7. Chigier, N.A., Mullinger, P.J.: The design and performance of internal mixing multijet twin fluid atomizers. J. Inst. Fuel **47**, 251–262 (1974)
8. Faeth, G.M., Hsiang, l.-P., Wu, P.-K.: Structure and breakup properties of sprays. Int. J. Multiphase Flow **21**(Suppl.) 99–127 (1995)
9. Schlottke, J., Weigand, B.: Direct numerical simulation of evaporating droplets. J. Comp. Phys. **227**, 5215–5237 (2008)
10. Spalding, D.B.: The combustion of liquid fuels. Proc. Combust. Inst. **4**, 847–864 (1953)
11. Godsave, G.A.E.: Studies of the combustion of drops in a fuel spray – the burning of single drops of fuel. Proc. Combust. Inst. **4**, 818–830 (1953)
12. Amsden, A.A.: KIVA-3V: A block-structured KIVA program for engines with vertical or canted valves, LA-13313-MS, UC-1412, Los Alamos National Laboratory, Los Alamos, NM (1977)
13. Williams, F.A.: Combustion Theory, 2nd edn. Benjamin/Cummings, Menlo Park (1985)
14. Sirignano, W.A.: Fluid Dynamics and Transport of Droplets and Sprays. Cambridge University Press, Cambridge (1999)
15. Williams, A.: Combustion of droplets of liquid fuels: a review. Combust. Flame **21**, 1–31 (1973)
16. Hubbard, G.L., Denny, V.E., Mills, A.F.: Droplet evaporation: effects of transient and variable properties. Int. J. Heat Mass Tran. **18**(9), 1003–1008 (1975)
17. Marchese, A.J., Dryer, F.L.: Effect of liquid mass transport on the combustion and extinction of bi component liquid droplets of methanol and water. Combust. Flame **105**, 104–122 (1996)
18. Aouina, Y., Gutheil, E., Maas, U., Riedel, U., Warnatz, J.: Mathematical modeling of droplet heating, vaporization, and ignition including detailed chemistry. Combust. Sci. Tech. **173**, 1–29 (2001)
19. Strauch, R., Lipp, S., Maas, U.: Detailed numerical simulations of the autoignition of single n-heptane droplets in air. Combust. Flame **145**(3), 533–542 (2006)
20. Stauch, R., Maas, U.: The ignition of methanol droplets in a laminar convective environment. Combust. Flame **153**(1–2), 45–57 (2008)
21. Düwel, I., Schorr, J., Wolfrum, J., Schulz, C.: Laser-induced fluorescence of tracers dissolved in evaporating droplets. Appl. Phys. B **78**, 127–131 (2004)
22. Ge, H.W., Urzica, D., Vogelgesang, M., Gutheil, E.: Modeling and simulation of turbulent non-reacting and reacting spray flows. In: Jäger, W., Rannacher, R., Warnatz, J. (eds.) Reactive Flows, Diffusion and Transport. Springer, Berlin (2006)
23. Urzica, D., Düwel, I., Schulz, C., Gutheil, E.: Laser-induced evaporation of a single droplet – An experimental and computational investigation. In: Proceedings of the 20th Annual Conference of ILASS, Europe, pp. 241–246 (2005)
24. Aulisa, E., Manservisi, S., Scardovelli, R., Zaleski, S.: A geometrical area-preserving volume-of-fluid method. J. Comput. Phys. **192**, 355–364 (2003)
25. Fu, W.B., Hou, L.Y., Wang, L., Ma, F.H.: A unified model for the micro-explosion of emulsified droplets of oil and water. Fuel Process. Technol. **79**(2), 107–119 (2002)
26. Shabde, V.S., Emets, S.V., Mann, U., Hoo, K.A., Carlson, N.N., Gladysz, G.M.: Modeling a hollow micro-particle production process. Comput. Chem. Eng. **29**, 2420–2428 (2005)
27. Kim, I., Elghobashi, S.E., Sirignano, W.A.: Three-dimensional flow over two spheres placed side-by-side. J. Fluid Mech. **246**, 465–468 (1993)
28. Srinivas, Y., Gutheil, E.: Modelling of droplet interaction in convective gas flows. Prog. Computat. Fluid Dyn. **4**(3–5), 250–256 (2004)

29. Kaltz, T.L., Long, L.N., Micc, M.M., Little, J.K.: Supercritical vaporization of liquid oxygen droplets using molecular dynamics. Combust. Sci. Tech. **136**, 279–301 (1998)
30. Malevanets, A., Kapral, R.: Mesoscopic, multi-particle collision model for fluid flow and molecular dynamics. Lecture Notes in Physics, vol. 640, pp. 116–149 (2004)
31. Tseng, L.-K., Ruff, G.A., Faeth, G.M.: Effects of gas density on the structure of liquid jets in still gases. AIAA J. **30**(6), 1537–1544 (1992)
32. Faeth, G.M.: Spray combustion phenomena. Proc. Combust. Inst. **26**, 1593–1612 (1996)
33. Vujanović, M., Edelbauer, W., von Berg, E., Tatschl, R., Duić, N.: Enhancement and validation of an Eulerian-Eulerian approach for diesel sprays. In: Proceedings of the ILASS Europe Conference, Paper 2–4, Como (2008)
34. Bayoro, F., Habachi, C., Daniel, E.: Numerical and physical basis of an Eulerian multi-phase flow model for the simulation of the liquid injection in internal combustion engines. In: Proceedings of the ILASS Europe Conference, Paper 2–5, Como (2008)
35. Drew, D.A., Passman, L.A.: Theory of Multicomponent Fluids. Springer, Berlin (1999)
36. Herrmann, M.: A Eulerian level set/vortex sheet method for two-phase interface dynamics. J. Comput. Phys. **203**, 539–571 (2005)
37. Sussman, M., Fatemi, E.: An efficient, interface-preserving level set redistancing algorithm and its application to interfacial incompressible fluid flow. SIAM J. Sci. Comput. **20**(4), 1165–1191 (1999)
38. Enright, D., Fedkiw, R., Ferziger, J., Michell, I.: A hybrid particle level set method for improved interface capturing. J. Comput. Phys. **148**, 2–22 (2002)
39. Stiesch, G.: Modeling Engine Spray and Combustion Processes (Heat and Mass Transfer). Springer, Berlin (2003)
40. Hollmann, C., Gutheil, E.: Modeling of turbulent spray diffusion flames including detailed chemistry. Proc. Combust. Inst. **26**(1), 1731–1738 (1996)
41. McDonell, V.G., Samuelsen, G.S.: An experimental data base for the computational fluid dynamics of reacting and nonreacting methanol sprays. J. Fluids Eng. **117**, 145–153 (1995)
42. Ge, H.-W., Düwel, I., Kronemayer, H., Dibble, R.W., Gutheil, E., Schulz, C., Wolfrum, J.: Laser based experimental and Monte Carlo PDF numerical investigation of an ethanol/air spray flame. Combust. Sci. Technol. **180**, 1529–1547 (2008)
43. Laurent, F., Massot, M.: Multi-fluid modelling of laminar polydisperse spray flames: origin, assumptions and comparison of sectional and sampling methods. Combust. Theory Modell. **5**, 537–572 (2001)
44. Laurent, F., Massot, M., Villedieu, P.: Eulerian multi-fluid modeling for the numerical simulation of coalescence in polydisperse dense liquid sprays. J. Comput. Phys. **194**, 505–543 (2004)
45. Williams, F.A.: Spray combustion and atomization. Phys. Fluids. **1**, 541–545 (1958)
46. Marchisio, D.L., Fox, R.O.: Solution of population balance equations using the direct quadrature method of moments. Aerosol Sci. **36**, 43–73 (2005)
47. Vikas, V., Wang, Z.J., Fox, R.O., Passalacqua, A.: A fully coupled fluid-particle flow solver using a quadrature-based moment method with high-order realizable schemes on unstructured grids. In: Proceedings of the International Conference on Multiphase Flow, Florida (2010).
48. Fox, R.O., Laurent, F., Massot, M.: Numerical simulation of spray coalescence in an Eulerian framework: direct quadrature method of moments and multi-fluid method. J. Comput. Phys. **227**(6), 3058–3088 (2008)
49. Rigopoulos, S.: Population balance modelling of polydispersed particles in reactive flows. Prog. Energy Combust. Sci. **36**, 412–443 (2010)
50. Réveillon, J.: Direct numerical simulation of sprays: turbulent dispersion, evaporation and combustion. In: Marchisio, D.L., Fox, R.O. (eds.) Multiphase Reacting Flows: Modelling and Simulation. Springer, New York (2007)
51. Wang, Y., Rutland, C.J.: Direct numerical simulation of ignition in turbulent n-heptane liquid-fuel spray jets. Combust. Flame **149**(4), 353–365 (2007)
52. Xia, J., Luo, K.H.: Direct numerical simulation of diluted combustion by evaporating droplets. Proc. Comb. Inst. **32**, 2267–2274 (2009)

53. Apte, S.V., Mahesh, K., Moin, P.: Large-eddy simulation of evaporating spray in a coaxial combustor. Proc. Comb. Inst. **32**, 2247 (2009)
54. Derksen, J.J.: Lattice-Boltzmann method for multiphase fluid flow simulations and Euler-Lagrange Large-Eddy simulations. In: Marchisio, D.L., Fox, R.O. (eds.) Multiphase Reacting Flows: Modelling and Simulation. Springer, New York (2007)
55. Boileau, M., Pascaud, S., Riber, E., Cuenot, B., Gicquel, L.Y.M., Poinsot, T.J., Cazalens, M.: Investigation of two-fluid methods for large eddy simulation of spray combustion in gas turbines. Flow Turbul. Combust. **80**(3), 291–321 (2008)
56. Subramanian, G., Colin, O., Pires da Cruz, A., Vervisch, L., Bruneaux, G.: Modeling turbulent mixing of an evaporating spray. In: Proceedings of the ICDERS, Montréal (2005)
57. Demoulin, F.X., Borghi, R.: Modeling of turbulent spray combustion with application to diesel like experiment. Combust. Flame **129**, 281–293 (2002)
58. Réveillon, J., Vervisch, L.: Spray vaporization in nonpremixed turbulent combustion modeling: a single droplet model. Combust. Flame **121**, 75–90 (2000)
59. Launder, B.E.: On the effects of a gravitational field on the turbulent transport of heat and momentum. J. Fluid Mech. **67**(3), 569 (1975)
60. Landenfeld, L., Kremer, A., Hassel, E., Janicka, J.: Comparison of Reynolds stress closures for strongly swirling combusting jets. In: 11th Symposium on Turbulent Shear Flows, Grenoble (1997)
61. Beishuizen, N.: PDF modeling and particle-turbulence interaction of turbulent spray flames. Ph.D. thesis, TU Delft, The Netherlands (2007)
62. Libby, P.A., Williams, F.A.: Turbulent Reactive Flows. Academic, London (1994)
63. Vogelgesang, M.: Reynolds stress modeling for turbulent spray flames. Ph.D. thesis (in German), Heidelberg University (2005)
64. Daly, H.J., Harlow, H.: Transport equations in turbulence. Phys. Fluids **13**, 2634–2649 (1970)
65. Gibson, M.M., Younis, B.A.: Calculation of swirling jets with a Reynolds stress closure. Phys. Fluids **29**(1), 38 (1986)
66. Launder, B.E., Reece, G.J., Rodi, W.: Progress in the development of a Reynolds-stress turbulent closure. J Fluid Mech. **68**(3), 537–566 (1975)
67. Hollmann, C., Gutheil, E.: Flamelet-modeling of turbulent spray diffusion flames based on a laminar spray flame library. Combust. Sci. Technol. **135**(1–6), 175 (1998)
68. Pope, S.N.B.: PDF methods for turbulent reactive flows. Prog. Energy Combust. Sci. **11**, 119–192 (1985)
69. Ge, H.-W., Gutheil, E.: PDF simulation of turbulent spray flows. Atomiz. Sprays **16**(5), 531–542 (2006)
70. Masri, A.R., Pope, S.N.B.: PDF calculations of piloted turbulent nonpremixed flames of methane. Combust. Flame **81**, 13–29 (1990)
71. Curl, R.L.: Dispersed phase mixing: I theory and effects in single reactors. AIChE J. **9**, 175–181 (1963)
72. Janicka, J., Kolbe, W., Kollmann, W.: Closure of the transport equation for the probability density function of turbulent scalar fields. J. Non-Equilib. Thermodyn. **4**, 47–66 (1979)
73. Masri, A.R., Subramaniam, S., Pope, S.N.B.: A mixing model to improve the PDF simulation of turbulent diffusion flames. Proc. Combust. Inst. **26**, 49–57 (1996)
74. Miller, R.S., Bellan, J.: On the validity of the assumed probability density function method for modeling binary mixing/reaction of evaporated vapor in gas-liquid turbulent shear flow. Proc. Combust. Inst. **27**, 1065–1072 (1998)
75. Ge, H.-W., Gutheil, E.: Simulation of a turbulent spray flame using coupled PDF gas phase and spray flamelet modeling. Combust. Flame **153**(1–2), 173–185 (2008)
76. Kee, R.J., Miller, J.A., Jefferson, T.H.: CHEMKIN: A general-purpose, problem-independent, transportable, Fortran chemical kinetics code package, Sandia Report SAN 80-8003, Sandia National Laboratories, Livermore, CA (1989)
77. Li, S.C.: Spray stagnation flames. Prog. Energy Combust. Sci. **23**(4), 303–347 (1997)
78. Continillo, G., Sirignano, W.A.: Counterflow spray combustion modelling. Combust. Flame **81**(3–4), 325–340 (1990)

79. Amantini, G., Frank, J.H., Smooke, M.D., Gomez, A.: Computational and experimental study of steady axisymmetric non-premixed methane counterflow flames. Combust. Theory Modell. **11**(1), 47–72 (2007)
80. Dakhlia, R.B., Giovangigli, V., Rosner, D.E.: Soret effects in laminar counterflow spray diffusion flames. Combust. Theory Modell. **6**(1), 1–7 (2002)
81. Gutheil, E., Sirignano, W.A.: Counterflow spray combustion modeling including detailed transport and detailed chemistry. Combust. Flame **113**(2), 92–105 (1998)
82. Schlotz, D., Gutheil, E.: Modeling of laminar mono- and bidisperse liquid oxygen/hydrogen spray flames in the counterflow configuration. Combust. Sci. Tech. **158**, 195–210 (2000)
83. Gutheil, E.: Structure and extinction of laminar ethanol/air spray flames. Combust. Theory Modell. **5**, 1–15 (2001)
84. Gutheil, E.: Multiple solutions for structures of laminar counterflow spray flames. Prog. Comput. Fluid Dyn. **5**(7), 414–419 (2005)
85. http://www.me.berkeley.edu/gri-mech/
86. Warnatz, J., Maas, U., Dibble, R.W.: Combustion. Springer, Berlin (1999)
87. Schlichting, H., Gertsen, K.: Boundary Layer Theory, 8th Revised and Enlarged Edition. Springer, Berlin Heidelberg New York (2000)
88. Peters, N.: Numerical and asymptotic analysis of systematically reduced reaction schemes for hydrocarbon flames. In: Glowinski, R., Larrouturou, B., Temam, R. (eds.) Numerical Simulation in Combustion Phenomena. Lecture Notes in Physics, vol. 241, pp. 90–109 (1985)
89. Peters, N., Rogg, B.: Reduced Kinetic Mechanisms for Applications in Combustion Systems. Springer, Berlin (1993)
90. Maas, U., Pope, S.B.: Simplifying chemical kinetics: intrinsic low-dimensional manifolds in composition space. Combust. Flame **88**, 239–264 (1992)
91. Peters, N.: Turbulent Combustion. Cambridge University Press, Cambridge (2000)
92. Bastiaans, R.J.M., Martin, S.M., Pitsch, H., van Oijen, J.A., de Goey, L.P.H.: Flamelet analysis of turbulent combustion. Lecture Notes in Computer Science, vol. 3516, pp. 64–71 (2005)
93. Peters, N.: Laminar flamelet concepts in turbulent combustion. Proc. Combust. Inst. **21**, 1231–1250 (1986)
94. Bastiaans, R.J.M., van Oijen, J.A., de Goey, L.P.H.: Application of flamelet generated manifolds and flamelet analysis of turbulent combustion. Int. J. Multiscale Comput. Eng. **4**(3), 307–317 (2006)
95. Bongers, H., van Oijen, J.A., Somers, L.M., de Goey, L.P.H.: The flamelet generated manifold method applied to steady planar partially premixed counterflow flames. Combust. Sci. Tech. **177**(12), 2373–2393 (2005)
96. Selle, L.C., Bellan, J.: Evaluation of assumed PDF methods in two-phase flows using direct numerical simulation. Proc. Comb. Inst. **31**, 2273–2281 (2007)
97. Klimenko, A.Y., Bilger, R.W.: Conditional moment closure for turbulent combustion. Prog. Energy Comb. Sci. **25**, 595–687 (1999)
98. Mortensen, M., Bilger, R.W.: Derivation of the conditional moment closure equations for spray combustion. Combust. Flame **156**, 62–72 (2009)
99. Merci, B., Mastorakos, E., Mura, A.: Modeling of turbulent combustion. In: Lackner, M., Winter, F., Agarwal, A.K. (eds.) Handbook of Combustion. Wiley-VCH, Weinheim (2010)
100. Wright, Y.M., de Paola, G., Boulouchos, K., Mastorakos, E.: Simulations of spray autoignition and flame establishment with two-dimensional CMC. Combust. Flame **143**, 402–419 (2005)

Chapter 2
Details and Complexities of Boundary Conditions in Turbulent Piloted Dilute Spray Jets and Flames

Assaad R. Masri and James D. Gounder

Abstract A good understanding of the details and sensitivities to boundary conditions is essential for the development of reliable predictive tools for turbulent spray flows. This chapter presents a comprehensive account of the boundary conditions relevant to dilute spray jets issuing in a co-flowing stream of air. The flames are stabilized with an annular pilot to prevent lift-off. Radial profiles of the velocity and turbulence fields as well as droplets size distributions and number densities are presented at the jet exit plane of a range of non-reacting and reacting jets of acetone and ethanol fuels. A non-reacting spray jet of mineral turpentine has been used as a base case to validate the liquid flux measurements. Laser-Phase Doppler Velocimetry (LDV/PDA) is employed for the measurements of velocity and droplet fluxes while Mie scattering is used for the high-speed laser imaging of droplets at the jet exit plane. The probability density function of the droplet size distribution measured at the exit plane is shown to be best represented by the Nukiyama-Tanisawa fit. Droplets less than 10 μm in diameter are shown to be adequate representatives of the gaseous flow. Larger droplets tend to exit the nozzle with mean axial velocities lower than those of the gas (and hence negative slip velocities). The rms fluctuations of the axial velocities of large droplets are found to be higher than those of small droplets. High-speed imaging of Mie scattering from sprays at the jet exit plane reveal that this anomaly is due to droplet shedding from the liquid film that develops on the inner wall of the pipe.

1 Introduction

The development of reliable numerical tools to aid in the design and optimization of combustors is now essential if demands for cleaner combustion systems are to be met. Such developments, while gradual in nature due to the multitude and

---

A.R. Masri (✉) and J.D. Gounder
School of Aerospace, Mechanical and Mechatronic Engineering, The University of Sydney, NSW, 2006 Australia
e-mail: assaad.masri@sydney.edu.au; James.Gounder@dlr.de

B. Merci et al. (eds.), *Experiments and Numerical Simulations of Diluted Spray Turbulent Combustion*, ERCOFTAC Series 17, DOI 10.1007/978-94-007-1409-0_2,

difficulties of the associated issues, have evolved significantly over the past decades particularly for gaseous flames. The International Workshop on Measurements and Calculations of Turbulent Non-premixed Flames (TNF) [1] was a key catalyst in this process bringing together modelers, experimental and numerical scientist to study generic flames that embody specific scientific issues such as turbulent-chemistry interactions [2–4], auto-ignition [5–7], and more recently highly sheared premixed combustion [8, 9]. Extensive data banks were made freely available to modelers and the difficulty of the problems tackled was gradually increased by introducing more complex flows involving recirculation and instabilities [10–12]. A clear outcome of this effort is a thorough understanding of turbulent gaseous combustion and a breakthrough in the capabilities of approaches such as the conditional moment closure [13, 14], flamelet/progress variable [15, 16] and the probability density function (PDF) methods [17, 18] to compute local extinction in turbulent non-premixed flames.

Spray flames have not enjoyed the same level of advance in computation capabilities as the gaseous counterparts despite their widespread use in applications such as diesel engines, turbines, rocket propulsion and liquid fueled furnaces. Admittedly, the issues here are far more complex with processes such as atomization, droplet coalescence and evaporation requiring modeling in conjunction with combustion. Interactions of droplets with the flowfield as well as with the heat release zones are additional factors that exacerbate these difficulties. Another key reason for the modest progress is the relative lack of detailed experimental data because the presence of droplets in the measurement probes limits the techniques that can be applied to such flows. Dense sprays are particularly prohibitive in this regard and require a level of sophistication that challenges current capabilities both in modeling as well as diagnostics. New measurements techniques, such as x-ray based methods [19] and ballistic methods [20] are being developed. Similarly, detailed computations of dense sprays resolving the complex process of atomization are emerging [21, 22].

Such difficulties led to a nominal separation of research activities into two broad schools, one studying dilute sprays and the other focusing on dense sprays including engines. Dense sprays generally dominate the regions close to the injector tip and involve transitions between a liquid core and an adjacent dispersed-flow region. Such flows are characterized by large liquid volume fractions; fluid filaments or irregular shaped droplets; collision, coalescence and potential breakup of liquid elements effects. The dispersed flow region generally begins at the injector exit for atomization breakup and consists of a developing multiphase mixing layer where the liquid core is present followed by a dispersed flow that eventually becomes a dilute spray. Dilute sprays refer to regions where the volume fraction of droplets is less than about 1% so that droplets are dispersed and may be assumed spherical in shape. Droplet collisions are infrequent in dilute sprays and heat and mass transfer and drag coefficient of individual droplets are independent of neighboring drops. Quantitatively a dispersed flow can be considered to be dilute if the mean inter-droplet spacing is significantly larger than the arithmetic mean diameter D10. However, there is a two way coupling between the phases, since transport from droplets influence the structure of the continuous phase. Dilute flows still involve complex and vaguely understood phenomena such as evaporation of droplets, turbulent dispersion and turbulence modulation of the gas phase flow due to the presence of droplets.

Thorough and extensive reviews of earlier research in spray laden flows [23–26] show that these cover a broad range of configurations such as particle laden homogeneous turbulent flows; particle laden gas jets; non-evaporating and combusting sprays; particle laden liquid jets, and non-condensing and condensing bubbly jets. Jets loaded with glass beads, solid particles, mono dispersed spray particles and poly-dispersed spray particles have been used by a number of authors [27–33] to study turbulent dispersion of droplets and turbulence modulation of the gas phase by particles/droplets. These studies have outlined the relationship between droplet size and the turbulent dispersion of droplets. Ferrand et al. [27] have shown the effect referred to as 'two-way coupling' in a turbulent jet loaded with droplets as well as quantified turbulence attenuation by the dispersed phase. Gore and Crowe [34], Hetsroni [30], and Yuan and Michaelides [33], studied the mechanisms for turbulence modulation. Two mechanisms for turbulence enhancement and reduction have been identified; (i) energy dissipation due to acceleration of a particle is the mechanism responsible for reduction of turbulence intensity of the flow, and (ii) flow disturbance due to the motion of the particle, its effective wake and the vortices shed behind, is the mechanism that contributes towards turbulence enhancement.

Two parallel international activities similar in nature to the TNF for gaseous flames are now attempting to coordinate experimental and numerical advances in the areas of dilute and dense sprays. The development of well designed generic burners that embody specific themes for further investigations and the generation of adequate and comprehensive data banks are essential requirements for success in both areas. There is a need to develop advanced diagnostics tools to enable the measurements of key quantities such as temperature and mixture fraction. Sufficient details of the boundary conditions must be provided so that comparisons with measurements as well as comparisons between separate calculations are valid. This requirement is particularly relevant for spray flows where the interaction of droplets with the solid wall of the burner as well as with any carrier fluid can impact significantly on the characteristics of the flow.

This chapter focuses on dilute sprays and is aimed providing a detailed account of the intricacies associated with the boundary conditions. The burner employed here was designed and developed at the University of Sydney with the objective of establishing it as model problem for studying streaming dilute jets and flames so the boundary conditions are relatively simple and well defined [35, 36]. The resulting flows are experimentally tractable as well as numerically simple to enable isolation of effects such as turbulence-chemistry, turbulence-droplet interactions, and turbulence effects on evaporation in dilute turbulent non-reacting and reacting spray jets. The complexities of the atomization process is avoided here by mounting the nebulizers at least 20 diameters upstream of the nozzle exit plane where an ultrasonic nebulizer is used for spray generation [36]. The ultrasonic nebulizer produces larger droplets of Sauter mean diameter around 40 μm, more akin to what is seen in industrial applications.

Section 2 gives a detailed description of the spray burner, the experimental setup and the conditions selected for later studies including a base, non-reacting case that uses mineral turpentine to validate the flux measurements. Mineral turpentine has a high boiling point ranging from 150°C to 190°C and hence is representative of non-evaporating sprays. Section 3 presents extensive details of the boundary conditions including profiles of the mean velocities, droplet size distributions and number densities. Section 3 compares the structure of a non-reacting spray jet with a gaseous jet of similar density. Mean velocities and turbulence levels of droplets of various sizes ranges are compared to those of gas. Section 4 describes the complexities that arise from the droplet's interaction with the inner walls of the fuel pipe leading to the jet exit plane. This is referred to as the spray pipe anomaly.

2 Experimental

2.1 The Burner

The piloted spray burner shown in Fig. 2.1 [37] comprises a base, a contraction with a contraction ratio of 10:1 and a pilot flame holder. A dimensioned schematic of the spray burner is provided in Fig. 2.1b. The central jet nozzle diameter D is 10.5 mm. The outer diameter of the annulus is 25.0 mm and the lip thickness is 0.2 mm. The pilot flame holder is fixed 7.0 mm upstream of the nozzle exit and holds 72 holes concentrically aligned at 7.0, 9.0 and 11 mm radius from the centre and each row contains 24 holes with diameters 0.9, 1.0 and 1.1 mm respectively. A co-flowing air stream with an inner diameter of 104 mm surrounds the burner and the entire co-flow/burner assembly is mounted in a vertical wind tunnel which has a rectangular exit cross section of 290 × 290 mm. The co-flow and nozzle exit plane is located 59.0 mm downstream of the tunnel exit plane. The co-flowing air velocity in the shroud as well as the wind tunnel was matched throughout the experiments and kept constant at 4.5 m/s. A more complete description of the burner assembly is presented elsewhere [37].

Spray is generated using Sono-Tek cooperation ultrasonic nebulizer model number 8700-48. A broadband ultrasonic generator is used for delivering the high frequency electrical energy required to operate the nebulizer. The nebulizer is centered inside the burner using four sets of screws. The nebulizer head is 215 mm upstream of the jet exit plane. Ultrasonically generated droplets of the liquid fuel with zero initial momentum are entrained in the carrier stream through a contraction as shown in the inset of Fig. 2.1b. The contraction is located 15 mm downstream of the tip of the nebulizer head and has a tapered inlet with a 9.5 mm radius. This arrangement was found to be optimal in minimizing asymmetry in the spray profile at the jet exit plane.

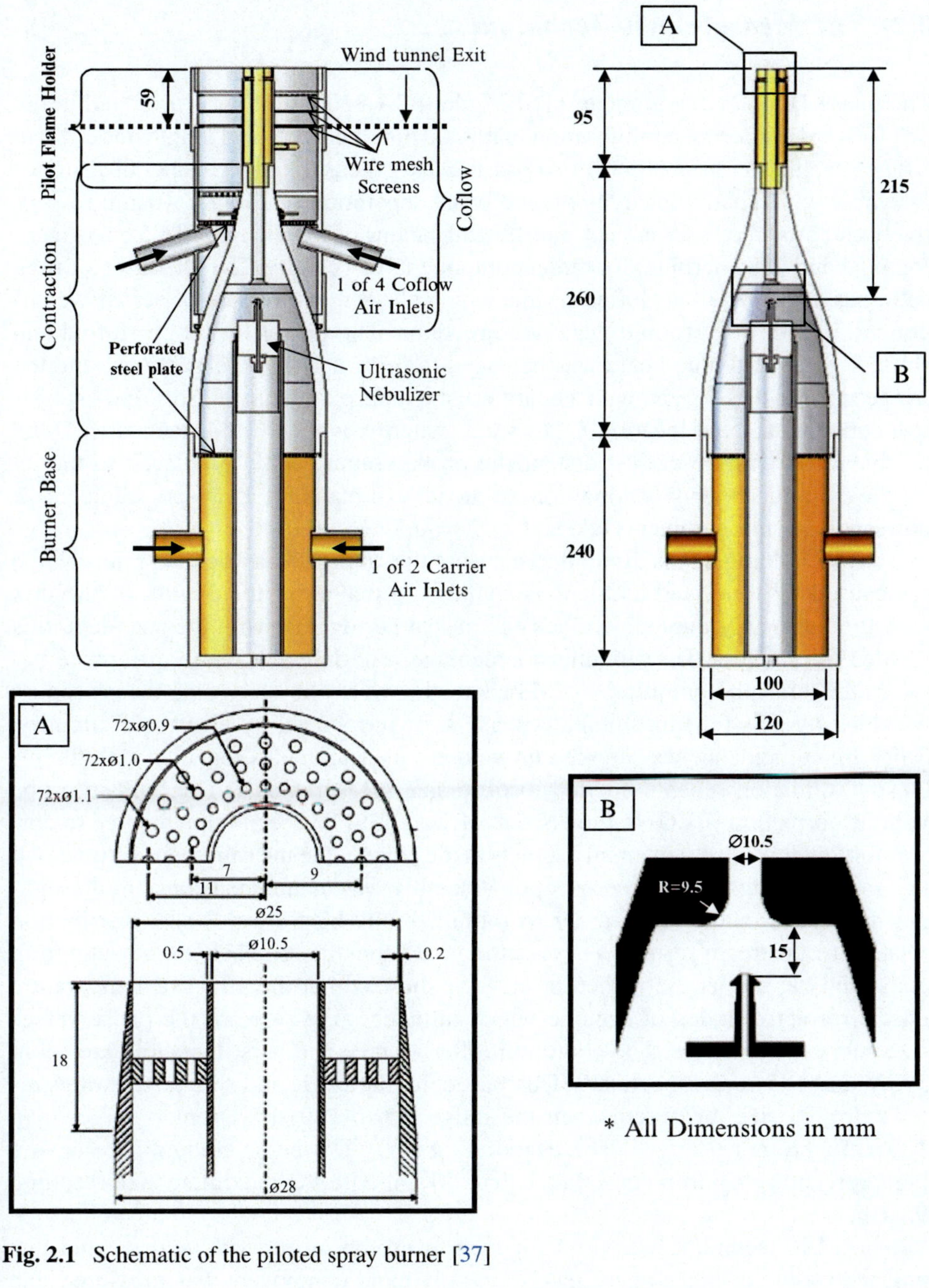

Fig. 2.1 Schematic of the piloted spray burner [37]

2.2 The Measurement Techniques

The phase-Doppler anemometer (TSI Model FSA 3500/4000) was arranged in the 45° forward scattering configuration, with 300 mm receiver focal length and 3.2 μm fringe spacing. A variable power Argon-ion laser feeds the two channel fiber optics assembly which had a focal length and beam separation of 250 and 40 mm respectively and produced two pairs of transmitting beams of wavelength 514.5 nm (green) for axial and 488 nm (blue) for radial components of velocity. The fiberlight's Bragg cell adds 40 MHz to the shifted beams which allow measurement of velocity component in both positive and negative directions. The power of the un-shifted and shifted, green and blue beam was balanced to 33 and 18 mW respectively and the fringe spacing was 3.2 μm with a beam waist of approximately 85 μm. Due to physical constraints, axial location $x/D = 0.3$ (where x is the axial distance and D the jet diameter) was the closest distant the probe volume could be placed to the jet nozzle exit, thus boundary conditions at the jet exit plane for spray jets studied here correspond to measurements taken at $x/D = 0.3$.

Droplet volume flux is difficult to measure due to high data rejection rates, which are caused by weak and ambiguous signals. To maximize the quality of the flux measurement, only the axial velocity is measured together with the size data, this yielded the highest size validation percentage. The inherent phase difference between the three photomultiplier (PMT) detectors was calibrated using the procedure provided by TSI [38] to eliminate phase shift introduced by electronic circuitry. Software coincidence was applied on velocity measurements together with the intensity validation scheme applied to drop size measurements. The built-in probe volume correction (PVC) in FlowSizer software [39] has been implemented to correct for lower visibility of small droplets at the edge of the measurement volume. All parameter settings of the receiver optic assembly was maintained constant throughout the length of the jet in order to minimize any bias towards size distribution relative to upstream results. These settings are thus determined by the saturation constraints at the jet exit. Chen et al. [35] showed that the effect of temperature on the refractive index of acetone which influences the slope of the phase verses size curve is negligible. The photomultiplier voltage and burst threshold are set at 395 V and 195 mV, respectively. This particular hardware and software settings allows droplet size measurement in the range 0.5 to 100 μm. Droplets larger than 100 μm were not resolved. The evaporating spray jets being studied here have a high percentage of droplets, smaller than 50 μm with the maximum size reaching 95 μm.

The LDV technique was used for measuring the mean axial velocity and turbulence profile of a gas phase jet. Note, only axial component was measured due to unavailability of the second channel since these measurements were performed later, after the second channel was damaged. The PMT voltage is set at 420 V and the burst threshold at 90 mV. The co-flow and gas phase jet stream was seeded with Aluminum oxide (Al_2O_3) particles of nominal diameter of 0.05 μm. The particles are entrained at a location upstream of the exit plane. Masri and Dally [40,41] have reported that there is uncertainty in these types of measurements due to velocity and seeding bias, however these errors are negligible if the velocity fluctuations are less than 10% which is the case for the jet studied here.

2.3 *The Blow-Off Limits*

The flames investigated here are generally steady and do not process. For a given fuel and atomizer, the parameters that could affect the stability limit of the spray flames are the mass flow rate of the main fuel and its carrier, the velocity and stoichiometry of the pilot and the velocity of the co-flowing air. Visible inspection of a spray flame for a fixed main fuel flow rate, pilot stoichiometry and velocity and co-flow velocity has shown that increasing carrier flow rate causes the flame to thin out, and gradually change color to light blue. Further increases in jet velocity causes a slight rumbling noise and the flame becomes visibly and intermittently disconnected in the region downstream of the pilot flame and this state of the flame has been defined as global blow off. Extensive tests have shown that the main controlling parameter in these flames is the carrier mass flow rate in the main fuel stream and the droplet loading. Stability limits for various fuels are presented elsewhere [42] and are reproduced here for convenience. Figure 2.2 shows the blow off limits of acetone and ethanol flames plotted against liquid fuel mass flow rate injected through the nebulizer. The broken lines represent the jet velocity (U_{jet}) and the solid lines the mass flow rate of the carrier at blow off. Two lines shown for U_{jet} and mass flow rate of carrier are the extreme cases chosen from a number of separate experiments. These lines in each of the plots reflect the level of measurement uncertainty associated with identifying these limits. The limits are shown for acetone and ethanol, in the respective figures plotted for a fixed co-flowing air velocity of 4.5 m/s, pilot unburnt velocity of 1.5 m/s and pilot equivalence ratio of 1.0.

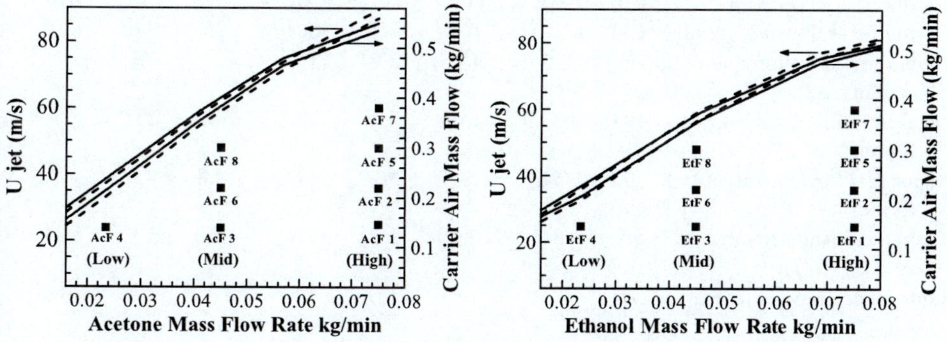

Fig. 2.2 Blow off limits for acetone and ethanol spray flames with air as carrier fluid. The *solid* and *dashed lines* show two different sets of measurements of bulk jet velocity and mass flow rate of the air carrier. Also shown are the flames selected for further measurements [37]

2.4 The Selected Cases

A total of 24 cases were selected for further study as detailed in Tables 2.1–2.3. Eight cases correspond to non-reacting jets of acetone with air used as carrier. These are referred to as SP1 to SP8 and relevant information about the liquid and vapor flowrates are given in Table 2.1. Eight flame conditions have been chosen for

Table 2.1 Relevant data and initial conditions for the selected non-reacting acetone jets

Non-reacting spray cases – acetone	SP1	SP2	SP3	SP4	SP5	SP6	SP7	SP8
Bulk jet velocity U_{jet} (m/s)	24	36	24	24	48	36	60	48
Carrier	Air	Air	Air	Air	Air	Air	Air	Air
Carrier mass flow rate (g/min)	150	225	150	150	301	225	376	301
Liquid fuel injection rate (g/min)	75	75	45	23.4	75	45	75	45
Measured liquid flow at exit (g/min)	22.1	28.8	16.7	10. 6	38.4	28.5	34.0	29.3
Vapor fuel flow rate at jet exit (g/min)	52.9	46.2	28.3	12.8	36.6	16.5	41.0	15.7
Spray jet density (kg/m^3)	1.52	1.44	1.42	1.33	1.40	1.38	1.36	1.34
Spray jet vapor phase viscosity (10^{-5}) (kg/m-s)	1.58	1.71	1.73	1.86	1.80	1.87	1.83	1.90
Jet Reynolds number	24,288	31,935	20,640	18,029	39,254	27,937	46,854	35,375

Table 2.2 Relevant data and initial conditions for the selected acetone flames

Reacting spray cases – acetone	AcF1	AcF2	AcF3	AcF4	AcF5	AcF6	AcF7	AcF8
Bulk jet velocity U_{jet} (m/s)	24	36	24	24	48	36	60	48
Carrier	Air	Air	Air	Air	Air	Air	Air	Air
Carrier mass flow rate (g/min)	150	225	150	150	301	225	376	301
Liquid fuel injection rate (g/min)	75	75	45	23.4	75	45	75	45
Measured liquid flow at exit (g/min)	18.0	23.9	15.9	7.2	27.8	26.2	31.1	27.4
Vapor fuel flow rate at jet exit (g/min)	57.0	51.1	29.1	16.2	47.2	18.8	43.9	17.6
Overall equivalence ratio ($\Phi_{overall}$)	4.7	3.2	2.9	1.5	2.4	1.9	1.9	1.4
Equivalence ratio at jet exit (Φ_{exitl})	3.6	2.2	1.8	1.0	1.5	0.8	1.1	0.6
Spray jet density (kg/m^3)	1.51	1.44	1.42	1.31	1.39	1.38	1.36	1.34
Spray jet vapor phase viscosity (10^{-5}) (kg/m-s)	1.56	1.69	1.73	1.83	1.77	1.86	1.82	1.90
Jet Reynolds number	24,417	32,131	20,730	18,103	39,609	28,076	47,076	35,526

Table 2.3 Relevant data and initial conditions for the selected ethanol flames

Reacting spray cases – Ethanol	EtF1	EtF2	EtF3	EtF4	EtF5	EtF6	EtF7	EtF8
Bulk jet velocity U_{jet} (m/s)	24	36	24	24	48	36	60	48
Carrier	Air	Air	Air	Air	Air	Air	Air	Air
Carrier mass flow rate (g/min)	150	225	150	150	301	225	376	301
Liquid fuel injection rate (g/min)	75	75	45	23.4	75	45	75	45
Measured liq. flow at exit (g/min)	45.7	66.6	30.7	14.5	70.1	41.3	73.0	36.6
Vapor fuel flow rate at jet exit (g/min)	29.3	8.4	14.3	8.9	4.9	3.7	2.0	8.4
Overall equivalence ratio ($\Phi_{overall}$)	4.7	3.2	2.9	1.5	2.4	1.9	1.9	1.4
Equivalence ratio at jet exit (Φ_{exitl})	1.75	0.34	0.85	0.53	0.15	0.15	0.05	0.25
Spray jet density (kg/m^3)	1.60	1.56	1.47	1.34	1.48	1.42	1.43	1.35
Spray jet vapor phase viscosity (10^{-5}) (kg/m-s)	1.79	1.92	1.88	1.92	1.96	1.96	1.97	1.96
Jet Reynolds number	22,525	30,661	19,678	17,506	38,190	27,422	45,735	34,848

each of acetone and ethanol fuels carried with air as shown in Tables 2.2 and 2.3 respectively. The acetone flames are referred to AcF1 to AcF8 while the ethanol flames are labeled EtF1 to EtF8. The eight flame conditions for the two fuels are shown on the stability plot in Fig. 2.2 for the three fuel loading cases with increasing carrier velocities. The relative liquid and gas flow rates were chosen to investigate the effects of increasing droplet loading at fixed carrier velocity (example: Cases 4, 3, 1 have increasing droplet loading) as well as the effects of varying carrier velocity for fixed liquid fuel flow rate (example: Cases 1, 2, 5, 7 have increasing carrier velocity). The three liquid fuel mass flow rates through the nebulizer were 23.4, 45 and 75 g/min and labeled as "Low", "Mid" and "High" respectively and the four carrier velocities were 24, 36, 48 and 60 m/s.

Relevant parameters for all the flames and the non-reacting jets are provided in Tables 2.1–2.3. Liquid fuel volume flow rate measured at the exit plane was obtained from phase Doppler particle anemometry measurements. The overall equivalence ratio $\Phi_{overall}$ of acetone and ethanol flames is calculated using initial mass flow rate of carrier air and the injected fuel mass flow rate assuming all of the liquid fuel is in vapor form. The equivalence ratio Φ_{exit}, at the jet exit plane is calculated using mass flow rate of fuel vapor measured at the exit plane and the initial carrier air flow rate. Fuel air mixtures at the exit plane in acetone flames are richer compared to the ethanol flames. This is due to larger amounts of acetone vaporizing inside the burner due to its lower boiling point at atmospheric pressure of 56°C, whereas ethanol has a boiling point of 78°C. The formulation for the spray jet density calculation is given in [42]. The gas phase temperature at the exit plane could not be measured, therefore it has been assumed to be 20°C for the density calculations. This is not true since there is significant amount of evaporation of fuel taking place inside the burner which would give lower gas phase temperature. The density calculated is an approximation thus the assumed gas phase temperature is valid. The viscosity of the spray jets were calculated using Chapman and Cowling [43] approximation for gas mixtures.

In addition to the non-reacting spray jets (SP1 to SP8), a gaseous jet has also been selected as a baseline for comparison with the spray jets. The gas issuing from the nozzle is chosen as a mixture of argon and nitrogen of equivalent density (1.4 kg/m^3) to the non-reacting acetone jet (case SP4 with a bulk jet velocity of 24 m/s) at the exit plane. Table 2.4 provides some relevant parameters for the gaseous jet baseline where the bulk wind tunnel and co-flow velocities were fixed at 4.5 m/s and the pilot stream has air flowing through at a bulk velocity of also 4.5 m/s. These conditions are maintained for consistency with the non-reacting jets so that comparisons of the velocity and rms fluctuations measured in the gaseous and non-reacting spray jets would give an indication of the effects of droplets on the flow and turbulence fields.

Normalised velocity and turbulence profiles at the exit plane of the gaseous jet and co-flow case (argon/nitrogen mixture) are shown in Fig. 2.3. In the central jet,

Table 2.4 Relevant data and initial conditions for the gas phase jet

	Argon/nitrogen mixture
Equivalent density spray case	SP4 (1.33 kg/m^3)
Bulk jet velocity (m/s)	24
Density (kg/m^3)	1.4
Viscosity (kg/m-s)	1.85×10^{-5}
Reynolds number	18,816

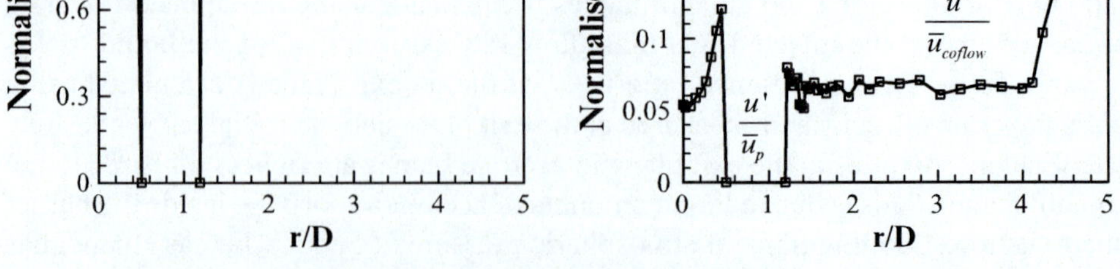

Fig. 2.3 Radial profiles of mean axial velocity and its rms fluctuations for gaseous flows at jet exit plane

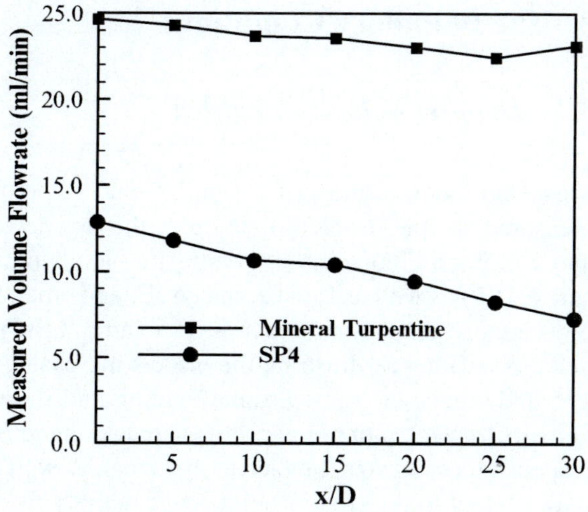

Fig. 2.4 Liquid volume flow rates measured at various axial locations in non-reacting spray jets of mineral turpentine and acetone

the measured mean velocity and its rms fluctuations are normalized with the bulk jet velocity, $\bar{u}_j = 24\,\text{m/s}$. The jet mean velocity and rms velocity profiles show a fully developed turbulent pipe flow at the exit plane of the nozzle. In the pilot annulus the exit plane condition was not measured but has been estimated as laminar flow with $\bar{u}_p = 4.5\,\text{m/s}$. Exit plane velocity of the co-flow stream was measured for $\bar{u}_{coflow} = 4.5\,\text{m/s}$. A uniform velocity profile is obtained at the exit plane of the co-flow with relative turbulence intensity of about 5%.

With its low boiling point of 56°C at atmospheric pressure, acetone is a good candidate for studying evaporation of sprays but not so good for validating the mass flux obtained from the LDV/PDA measurements. For this purpose, a much less volatile liquid is needed and mineral turpentine, with a boiling point around 150°C, is injected through the nebulizer at a bulk flow rate of 24.6 ml/min issuing at the jet exit plane with air as carrier. PDA Measurements were performed at various axial locations in this non-reacting jet and the droplet flow rate at each axial location was then obtained by integrating the measured droplet fluxes across the jet at the exit plane. The same experiment was repeated for acetone Case SP4, which has a liquid mass flow rate of 12.9 ml/min at the jet exit plane and the results for both acetone and mineral turpentine are shown in Fig. 2.4 in the form of measured volume flow rate plotted versus axial locations. Measurements in the mineral turpentine case recover over 90% of the injected liquid flow even at axial location as far downstream as $x/D = 30$, hence confirming the validity of the LDV/DPA measurements for droplet flux as well as velocities in an evaporating jet. For acetone, the significant reduction in the measured liquid flux with axial distance (over 50% at $x/D = 30$) is expected due to the high evaporation rate.

3 The Boundary Conditions

3.1 Droplet Velocity Profiles

Measured radial profiles of the mean axial velocity ($\bar{u}$) and it rms fluctuations (u') measured for the droplets flowing at the nozzle exit plane are shown in Figs. 2.5 and 2.6. Each Figure shows composite plots with results for selected non-reacting sprays (SP) as well as for acetone (AcF) and ethanol flames (EtF). For each fuel, six plots are presented for cases 4, 3, 1, 5 and 7. Cases 4, 3 and 1 have the same carrier velocity and hence illustrate the effects increasing the droplet loading while Cases 1, 5 and 7 have the same droplet loading and show the effects of increasing the jet velocity. In each plot radial profiles are presented for the results unconditioned with respect to droplet size (labeled as all sizes), as well as for measurements conditioned with respect to different droplet sizes, namely five specific droplet size bands: $d < 10$, $10 < d < 20$, $20 < d < 30$, $30 < d < 40$ and $40 < d < 50$, where d is the droplet diameter. Profiles corresponding to the smallest droplet size ($d < 10\,\mu m$) are most likely to correspond to the gas phase velocity. This aspect is confirmed later. It should be noted here that plots for the mean axial velocity have a maximum range of 35 m/s on the vertical axis for Cases 4, 3 and 1 increasing to 63 and 72 m/s for cases 5 and 7 respectively. The rms velocity plots have a fixed range of 7 m/s for spray jets with increasing fuel loading at constant carrier velocity (Cases 4, 3 and 1) and 12 m/s for fixed high droplet fuel loading with increasing carrier velocity cases (Cases 5 and 7). The horizontal axis represents radial distance (r) normalized by the jet nozzle diameter D.

The plots shown in Figs. 2.5 and 2.6 are very informative. Conditional mean axial velocities measured at the jet exit plane show consistent trends for the reacting (EtF, AcF) and non-reacting (SP) flows as well as for variations in the droplet loading (Cases 4, 3, 1) or the fuel jet velocity (Cases 1, 5, 7). Large droplets have lower mean axial velocities than smaller droplets on the centerline with the difference decreasing as the tube wall is approached. Assuming the gas phase velocity is represented by that of droplets with $d < 10\,\mu m$, then large droplets lag initially behind the gas phase (small droplets) due to high inertia and thus exit the nozzle with a negative slip velocity $\bar{u}_s$ given by:

$$\bar{u}_s = \bar{u} - \bar{u}_g \tag{2.1}$$

where $\bar{u}$ is the mean velocity of the droplet and $\bar{u}_g$ is the gas phase velocity represented by velocity of droplets smaller than $10\,\mu m$. The magnitude of slip velocity increases with droplet size. Only a small change in the slip velocity of the large particles is observed as the fuel loading is increased. In spray cases with fixed fuel loading and increasing carrier velocity, the magnitude of the slip velocity of large droplets increases as gas phase velocity increases.

Radial profiles of axial turbulence (Fig. 2.6) show consistent trends of low rms fluctuations on the centerline with increasing towards the wall of the pipe and

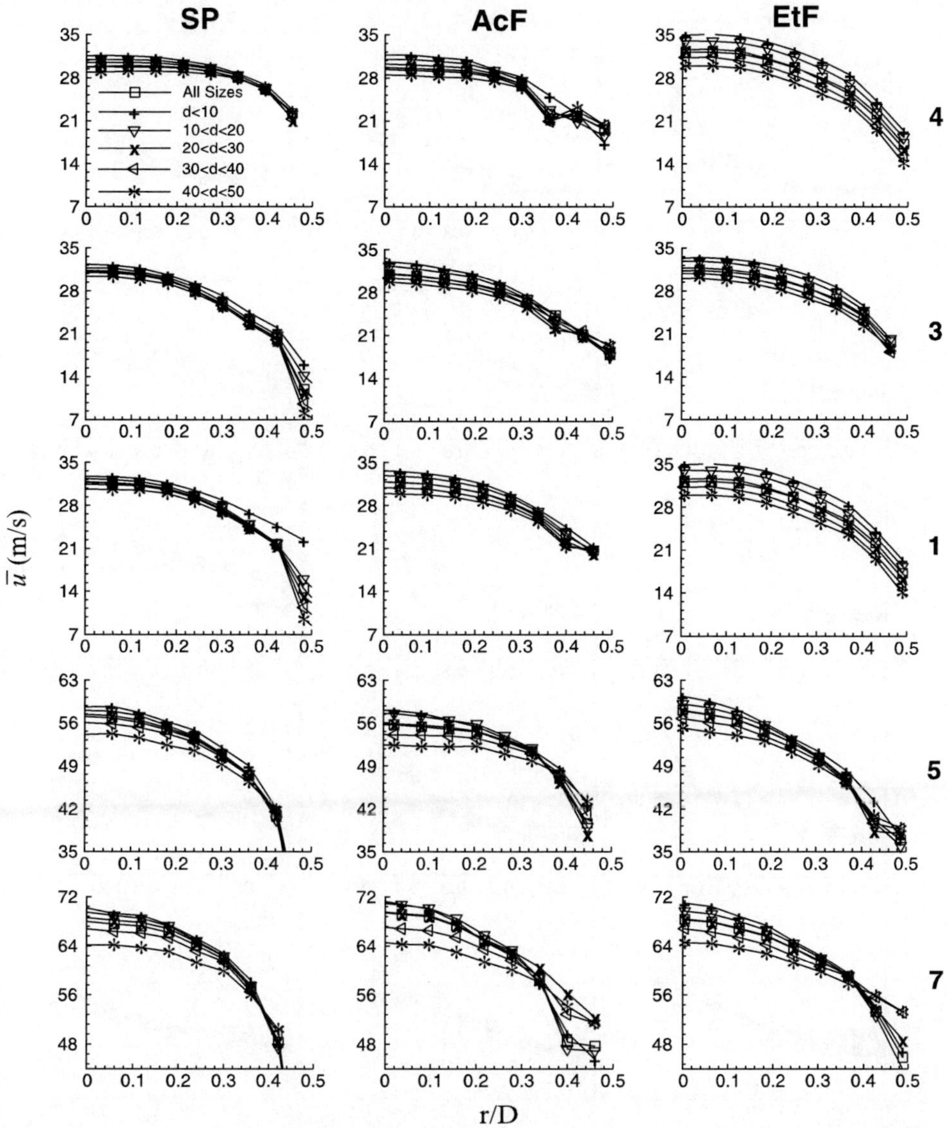

Fig. 2.5 Radial profiles of mean axial velocity $\bar{u}$ measured at the exit plane (x/D = 0.3) for cases with constant carrier velocity and increasing fuel loading (4, 3, 1) and constant fuel loading (High) and increasing carrier velocity (1, 5, 7) in non-reacting spray jets (SP) and reacting spray flames of acetone (AcF) and ethanol (EtF). Results are presented for unconditioned case (all sizes) as well as conditioned with respect to five droplet size bins ($d < 10$, $10 < d < 20$), $20 < d < 30$, $30 < d < 40$ and $40 < d < 50$)

peaking either at the wall or just a small distance away from the wall. For the non-reacting cases (SP) and most of the reacting cases (except for EtF3, AcF3, EtF1 and AcF1), the rms fluctuations measured for all droplet sizes are similar on the centerline. However, as the pipe wall is approached, larger droplets start

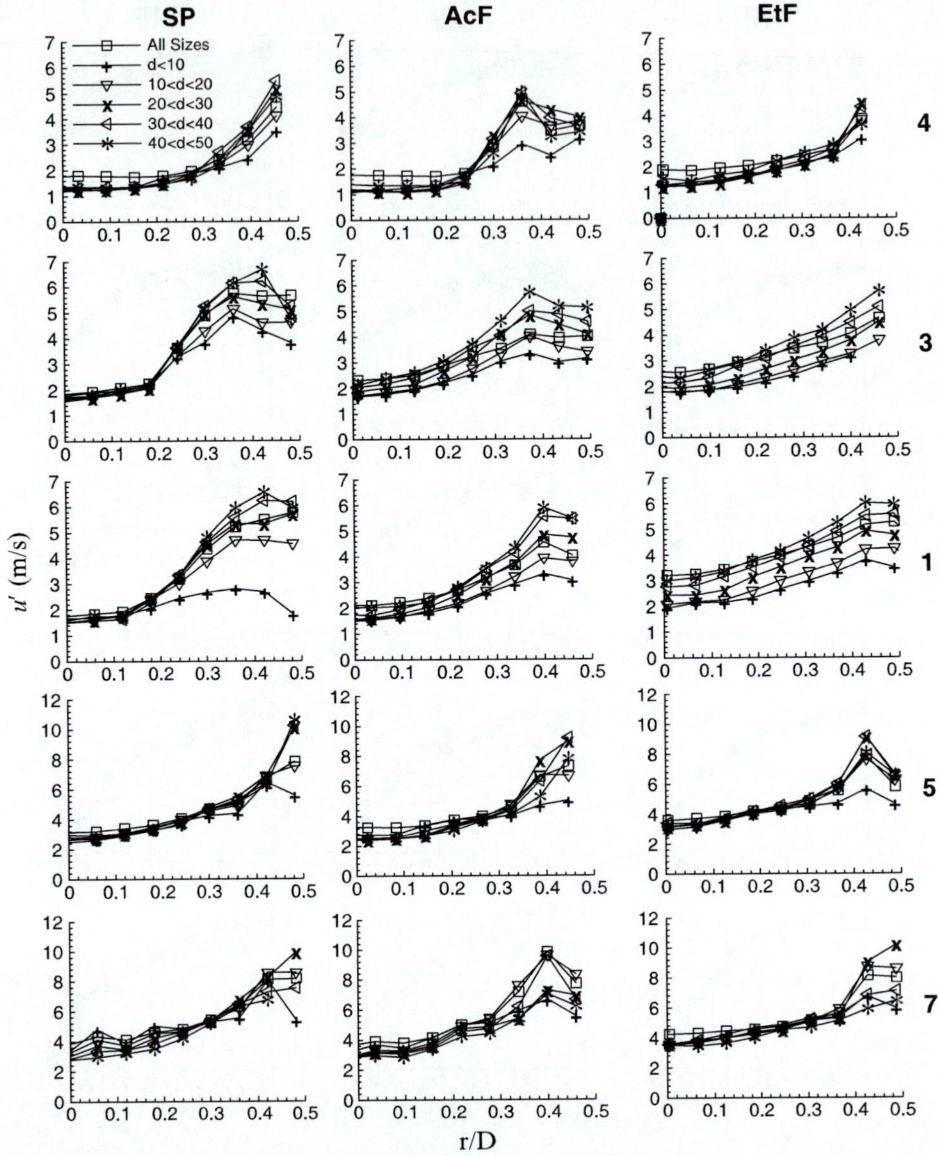

Fig. 2.6 Radial profiles of rms velocity fluctuations u' measured at the exit plane (x/D = 0.3) for cases with constant carrier velocity and increasing fuel loading (4, 3, 1) and constant fuel loading (high) and increasing carrier velocity (1, 5, 7) in non-reacting spray jets (SP) and reacting spray flames of acetone (AcF) and ethanol (EtF). Results are presented for unconditioned case (all sizes) as well as conditioned with respect to five droplet size bins ($d < 10$, $10 < d < 20$), $20 < d < 30$, $30 < d < 40$ and $40 < d < 50$)

to show higher rms fluctuations. This is consistent with the finding of Starner et al. [36] in turbulent spray flames of acetone, who reported rms fluctuations for the axial velocity measured for large droplets near the wall at the jet exit plane higher than those measured for droplets smaller than 10 μm. This finding is contrary to expectations that small droplets with lower inertia are more likely to follow the turbulent fluctuations of the gas phase and show higher rms values. At larger diameters, the inertia of the droplets increases and hence these maintain their speed but are less likely to respond to flow turbulence fluctuations. The higher rms fluctuations measured for larger droplets are, therefore, anomalous and this is referred to as the spray pipe anomaly which is addressed in Sect. 5 of this chapter.

3.2 Droplet Size Distribution

Droplet size distribution measured close to the jet center at the jet exit plane ($x/D = 0.3$) are presented in Fig. 2.7 in the form of probability density functions for selected cases in the non-reacting as well as the reacting acetone and ethanol sprays. The effects of increasing the spray loading for a given jet velocity are shown in plots on the left hand side for Cases 1, 3 and 4 while the effects of increasing the jet Reynolds number for a given droplet loading are shown on the right hand side for Cases 1, 2, 5 and 7. The measured droplet probability density function (PDF) has been normalized by the maximum diameter count of that particular spray case.

For flows with constant carrier velocity but increasing droplet loading (Cases 4, 3, 1 shown on the LHS of Fig. 2.7), both the non-reacting and reacting jets show somewhat similar droplet size distributions with the peak of the PDF lying between droplet diameters of 10 to 40 μm. The only outliers here are the acetone flames which show a gradually narrowing distribution for the more heavily loaded Cases, AcF3 and AcF1. For a fixed droplet loading, increasing carrier velocity leads to a decrease in the peak of the PDF distribution to a lower droplet size. This is consistent for the non-reacting as well as the reacting cases of both acetone and ethanol fuels. Increasing the jet velocity from Cases 1 to 7 shifts the PDF distribution by at least 10 μm towards smaller droplets. This implies that higher velocities cause higher evaporation rates for larger droplets hence shifting the overall distribution towards the smaller size range.

Evaporation of droplets into the carrier flow and the transit history in the pipe as the spray/carrier mixture is proceeding towards the jet exit plane are thought to be the main cause of the modified PDF's measured at the nozzle exit. Non-uniformities in the distribution due to interactions of the spray with the pipe walls are addressed Sect. 5 of this Chapter. The droplet size distribution produced by the ultrasonic nebulizer is known to be approximately lognormal. As the droplets evaporate and interact with the carrier fluid the shape of the original distribution changes slightly. It is found that the PDF's measured at the jet exit plane and shown in Fig. 2.7 are best represented by the Nukiyama-Tanasawa [44] droplet size distribution. Figure 2.8 shows a comparison between the measured droplet size PDF in non-reacting spray jet SP3 and Nukiyama-Tanasawa fit.

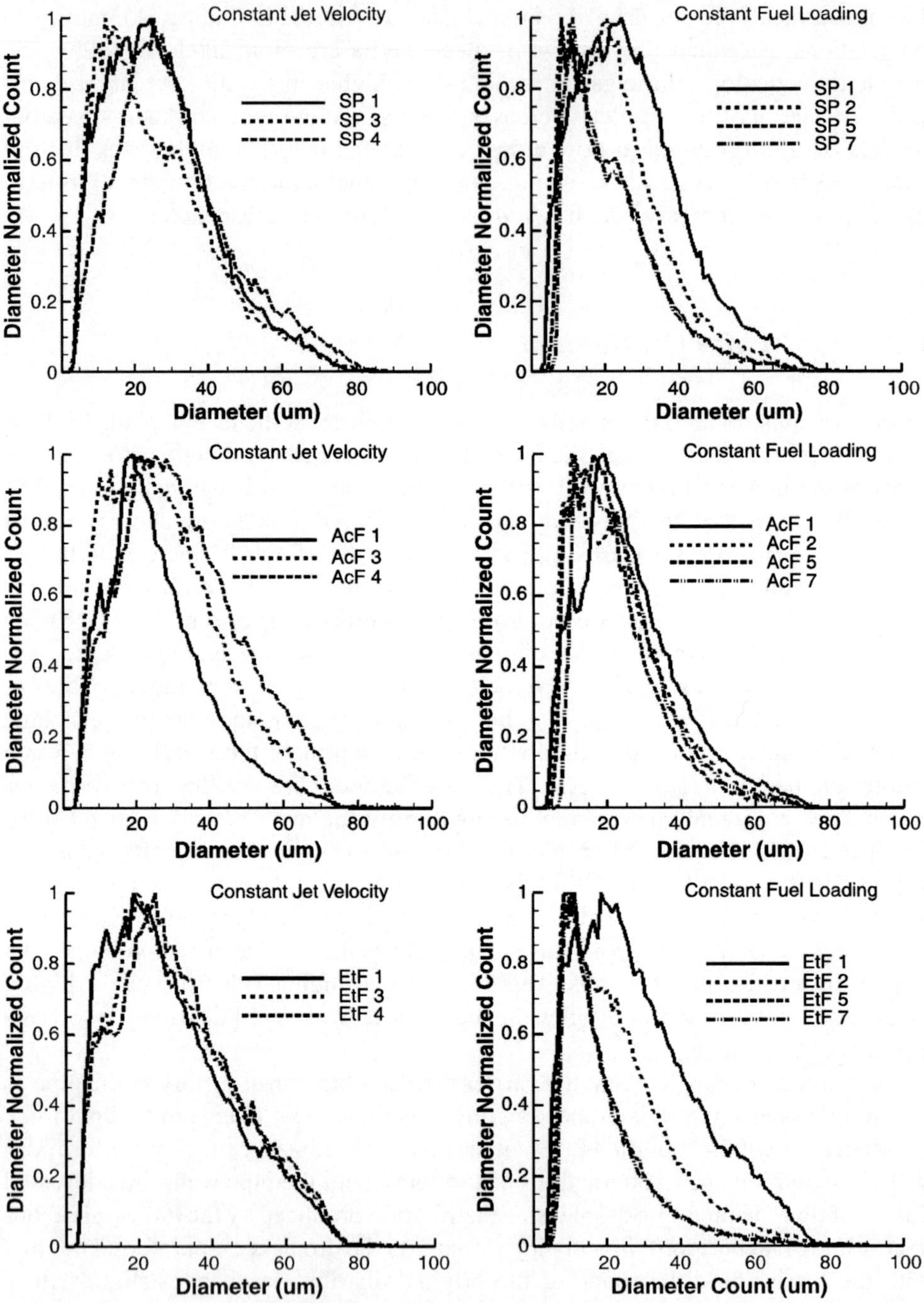

Fig. 2.7 Droplet size distribution close to the jet central axis at x/D = 0.3 for non-reacting spray jets, acetone and ethanol spray flames. First column contains size distribution for jets with increasing fuel flow rate and constant carrier velocity, second column contains drop size distribution for jets with fixed fuel flow rate and increasing carrier velocity

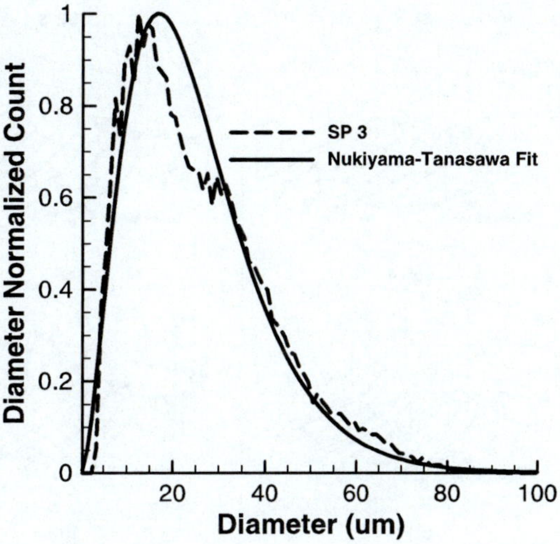

Fig. 2.8 Comparison between measured and theoretical droplet size distribution. Nukiyama-Tanisawa droplet size distribution provides the best fit to the measured droplet size distribution at the nozzle exit plane

Hinze [45] has shown that droplet deformation and secondary break up regimes are functions of Weber (We) and Ohnesorge (Oh) numbers. Weber number is defined as the ratio of dynamic force to surface-tension force and is given by:

$$We = \frac{d\rho_g |u_p - u_g|^2}{\sigma} \tag{2.2}$$

where d is the particle diameter, ρ_g is the gas density, u_p is the stream wise particle velocity, u_g is the stream wise gas velocity and σ is the surface-tension. Ohnesorge number on the other hand is the ratio of liquid viscous forces to surface tension forces and ven by:

$$Oh = \frac{\mu_l}{\sqrt{\rho_l d \sigma}} \tag{2.3}$$

where μ_l is the liquid viscosity, ρ_l is the liquid density, d is the particle diameter and σ is the surface tension. Weber number (We) and the Ohnesorge number Oh of the spray at the exit plane has been calculated using Eqs. 2.1 and 2.2 for all spray jets. The drop deformation regime map as a function of We and Oh, provided by Faeth [46] was used for determining if further droplet deformation or break up occurred in the spray jets investigated here. The Weber number was found to be less than 0.3 and Oh was less than 0.006 for droplets in the sizes range $40 < d < 50\,\mu m$ indicating that the droplets are spherical and will not undergo any further deformation due to droplet break up, downstream of the nozzle exit plane.

It should be noted that the droplet distributions shown in Figs. 2.7 and 2.8 are measured on the jet centerline as close as possible to the jet exit plane. Radial scans of the droplet size PDF at the exit plane are shown in Figs. 2.9 and 2.10 for the non-reacting (SP) as well as reacting jets of both acetone (AcF) and ethanol (EtF)

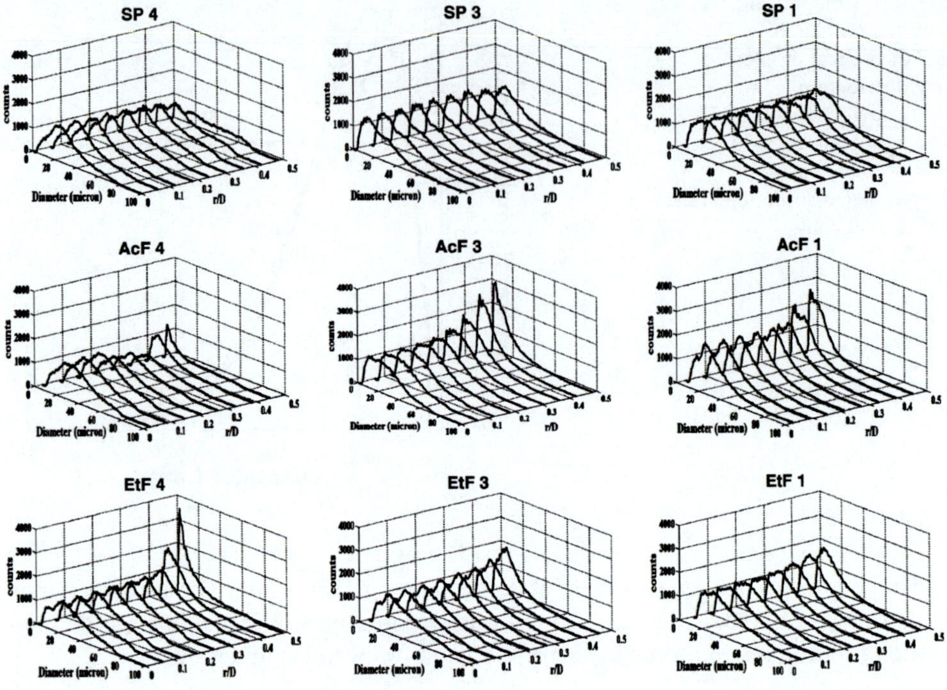

Fig. 2.9 Radial distribution of droplet size PDF's measured at the exit plane for cases with constant carrier velocity and increasing fuel loading for non-reacting spray jets (SP4, SP3, SP1), acetone (AcF4, AcF3, AcF1) and ethanol (EtF4, EtF3, EtF1) flames respectively [37]

fuels. Figure 2.9 shows measured PDF's for Cases 4, 3 and 1 which mark the effects of increasing the droplet loading for a fixed carrier velocity while Fig. 2.10 shows PDF's for Cases 2, 5 and 7 which show the effects of increasing the Reynolds number for a fixed droplet loading. Each plot in Figs. 2.9 and 2.10 shows a range of PDF's for the droplet distributions measured at various radial locations at the jet exit plane starting from the centerline at $r/D = 0$ and progressing towards the inner tube wall at $r/D = 0.5$.

Figures 2.9 and 2.10 show that along the radial profile of the jet exit plane, the droplet size distributions remain qualitatively similar to those measured on the centerline except near the pipe wall where the PDFs start to show a bias towards small droplets ($d = 10\,\mu m$). This trend is repeated in most of the 18 cases presented here. With increasing droplet loading, as shown in Cases 4, 3, and 1 presented in Fig. 2.9, the radial distribution of droplet size PDF's show little change with a peak around 20 μm except near the wall. The only exceptions here are the non-reacting jets SP4, SP3 and SP1 where the bias for small droplets expected near the wall is not observed. Figure 2.10 shows that, for a fixed droplet loading, increasing the jet velocity leads to a distinct transition in the peak of the droplet size PDF towards smaller droplets as the carrier velocity is increased. A consistent further shift toward smaller droplets is observed here for all cases as the jet wall is approached.

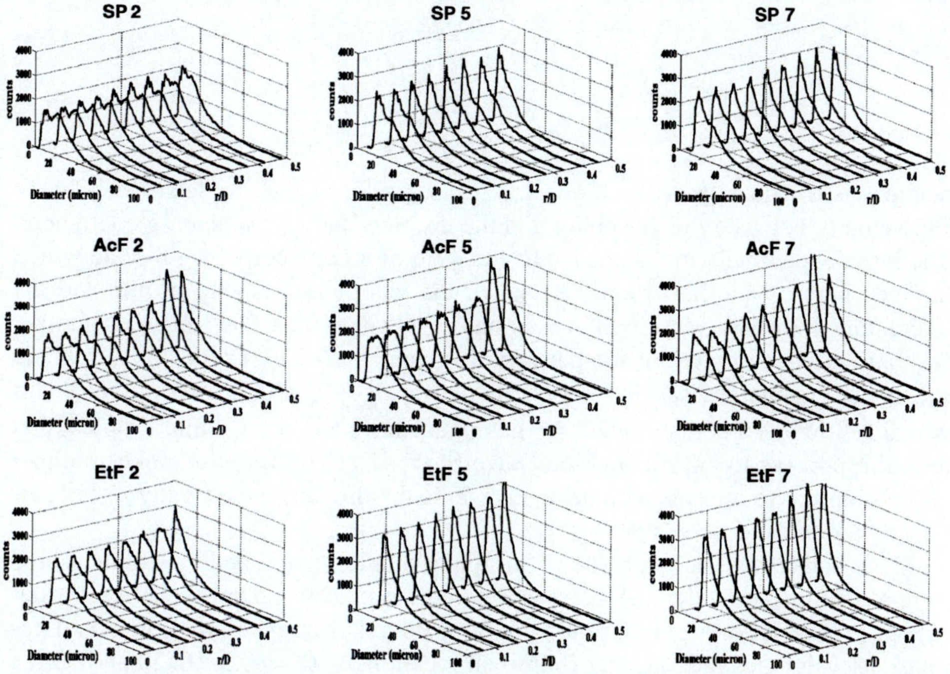

Fig. 2.10 Radial distribution of droplet size PDF's measured at the exit plane for cases with increasing carrier velocity and fixed fuel loading (high) for non-reacting spray jets (SP2, SP5, SP7), acetone (AcF2, AcF5, AcF7) and ethanol (EtF2, EtF5, EtF7) flames respectively [37]

4 Comparison with Gaseous Jets

In earlier work [36,47,48] it has been assumed that the gas phase flow field behavior can be represented by mean and rms velocities of droplets with diameter less than 10 μm. To confirm this assumption, profiles of the mean axial velocity and its rms fluctuations measured in a gaseous jet are compared to those measured from droplets in an evaporating spray jet. The initial conditions of the gas phase jet consisting of argon and nitrogen mixture are provided in Table 2.4. The bulk jet velocity of the argon/nitrogen mixture is matched to the bulk carrier velocity of the non-reacting spray case SP4 at 24 m/s. Similarly, the density of the argon/nitrogen mixture at the jet exit plane is also matched to that of the SP4 spray jet.

The experimental set up for velocity measurements in the gaseous jet is described above where 0.05 μm aluminum oxide particles seeded in the argon/nitrogen mixture provide a measure of the mean and rms velocity. The Stokes number (St) marks the rate at which particles or droplets respond to turbulent fluctuations and is defined as the ratio of the droplet relaxation time τ_d, and the turbulent time scale τ_t, which are given by following expressions

$$\tau_d = \frac{4}{3}\frac{\rho_{droplet}}{\rho_{gas}}\frac{d}{C_D\langle|\bar{u}_s|\rangle} \tag{2.4}$$

$$\tau_t = \frac{R_{1/2\bar{u}}}{u'} \tag{2.5}$$

$$St = \frac{\tau_d}{\tau_t} \tag{2.6}$$

where ρ is the density of the fluid, d is the diameter of the droplets, $\langle|\bar{u}_s|\rangle$ is the slip velocity between the gas phase and the droplets and C_D is the drag coefficient. The turbulence time scale is chosen as the ratio of a large eddy length scale which is given here by the half radius $R_{1/2\bar{u}}$, of the gas phase velocity profile and the fluctuating gas phase velocity u', taken on the jet centerline. The Reynolds number Re_p, and drag coefficient of the particles is given by $Re_p = (d\langle|\bar{u}_s|\rangle)/\nu_{gas}$ where ν_{gas} is the kinematic viscosity of gas, and $C_D = \frac{24}{Re_p}(1 + 0.15Re_p^{0.678})$. Particles which follow the fluctuations of the gas phase have St << 1 while droplets that are unresponsive to any fluctuations have St >> 1. Particles with stokes number close to unit have intermediate behavior where the droplets are partially responsive depending on the droplet size.

Mean and rms velocity profiles of droplets smaller than 5 and 10 μm measured in non-reacting spray case SP4, are compared with mean and rms velocity profiles of a gas phase jet of argon/nitrogen mixture. Radial profiles of the droplet and gas phase jet velocities measured at four axial locations $x/D = 0.3, 10, 20$ and 30 are presented in Fig. 2.11. At the exit plane ($x/D = 0.3$) the droplet velocity profiles match up nicely with the velocity profiles of the gaseous mixture. Profiles of ū for the argon/nitrogen mixture are lower than the droplet mean velocities in the central core of the jet at $x/D = 10$ and 20 however the u' profiles show very small difference. Further downstream at $x/D = 30$ velocity profiles of droplets and the argon/nitrogen mixture jet match up again. The difference in the mean velocity profile at $x/D = 10$ is believed to be due to the variation in the level of entrainment of co-flow air in the purely gaseous jet (argon/nitrogen mixture) and the evaporating spray jet. The level of entrainment and the intensity of mixing affect the dilution of the jet mixture thus changing the jet mixture density. In the gaseous mixture case, the jet density would reduce while the same cannot be said for the evaporating spray jet due to the presence of droplets, hence affecting mixing with the ambient air. The variation in the density of the jet mixture and the level of mixing could be the influencing factors responsible for the slow decay of mean velocity of droplet particles at $x/D = 10$ and 20. Therefore, variation in the mean profile could be due to any one of these factors. A final point to note is that there is very small difference between velocity profiles of droplets smaller than 5 μm and droplets smaller that 10 μm.

A comparison is made between Stokes numbers (St) calculated for droplets less than 10 μm based on two sets of assumptions for representation of turbulence fluctuations of gas phase in the spray jet. Firstly assuming that droplets smaller than 5 μm follow the turbulent fluctuations of the carrier air and thus the turbulence time scale is obtained from u' of droplets smaller than 5 μm. The second assumption is using turbulence fluctuations measured in the argon/nitrogen mixture to represent the carrier u' in the spray jet. The St calculated using $u' = u'_{d<5\,\mu m}$ range from 0.05 at $x/D = 0.3$ to 0.01 at $x/D = 30$. Similarly St calculated using $u' = u'_{Ar/N2mix}$ is

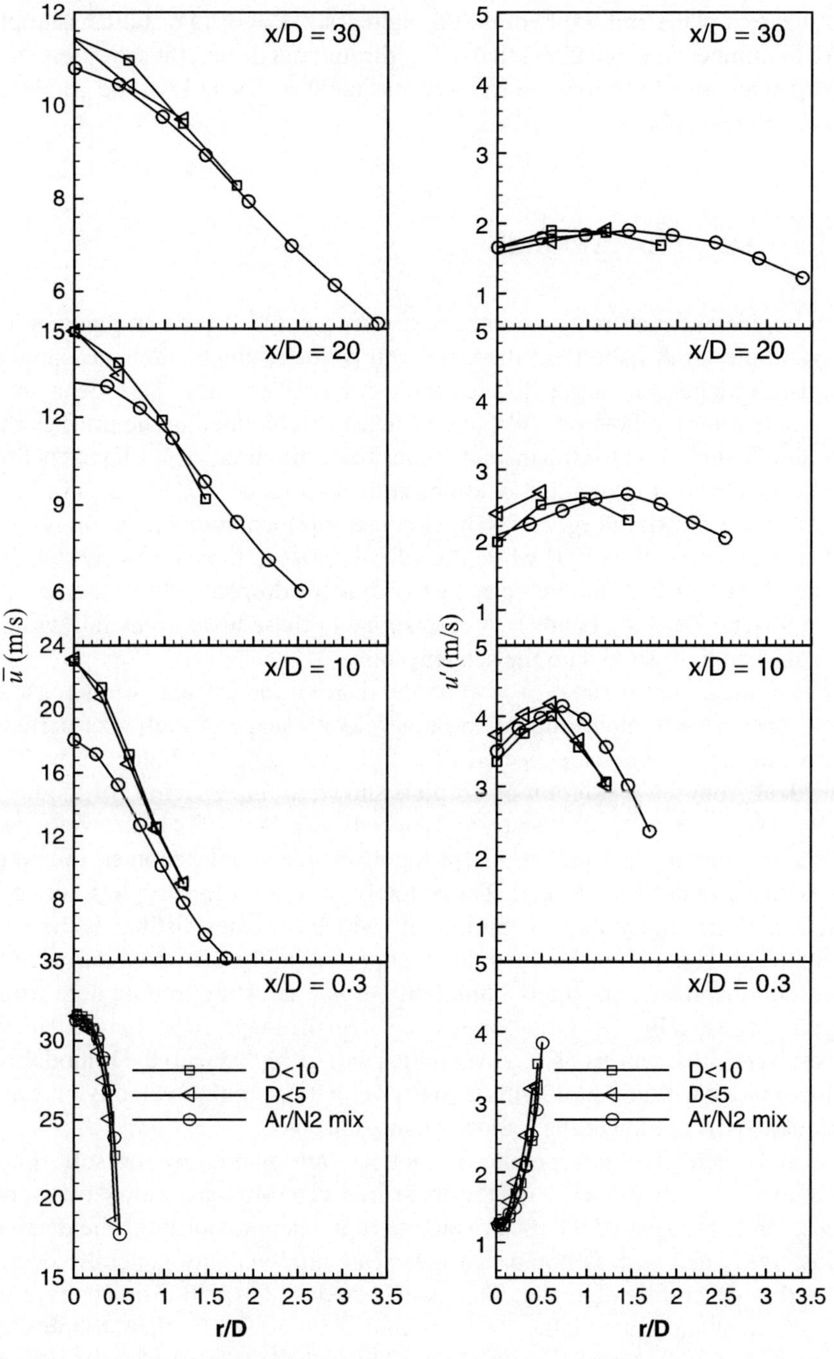

Fig. 2.11 Radial profiles of mean ($\bar{u}$) and rms (u') velocity of 5 and 10 mm droplets in spray case SP4 compared with velocity profiles of argon/nitrogen mixture at four axial locations (x/D = 0.3, 10, 20 and 30)

0.04 at the exit plane and 0.009 in the far field at $x/D = 30$. For both assumptions the Stokes numbers are much less than unity throughout the jet, thus droplets smaller than 10 μm are able to follow the smallest turbulent scales and will be good tracers for the continuous phase.

5 The Spray-Pipe Anomaly

The spray pipe anomaly refers to the unexpected finding at the exit plane of a pipe carrying dilute sprays showing the axial rms velocity fluctuation measured near the wall are higher for larger droplets than for smaller ones. This behavior was reported in earlier publications [49] and is clearly highlighted in the profiles shown in Fig. 2.6. Scatter plots for the instantaneous realizations of velocity versus droplet diameter measured at two radial locations at the exit plane ($x/D = 0.3$) of acetone flame AcF8 are shown in Fig. 2.12. The first measurement radial location is close to the jet centerline ($r/D = 0.05$) while the other is close to the jet wall ($r/D = 0.41$). Also shown on Fig. 2.12 are velocity histograms for droplets within the size range of 20 to 30 μm. The data bands that correspond to these histograms fall within the vertical dashed lines shown on the scatter plots.

Close to the centerline at $r/D = 0.05$ the distribution of data points for a given bin of droplet size is mono-modal and the velocity scatter for all size particles is close to the mean velocity expected of a fully developed turbulent flow. This is also evident from the histogram of droplets shown at this location for the droplets with $20 < d < 30\,\mu$m. Closer to the jet wall at $r/D = 0.41$, the velocity distribution is generally broader even for small droplets and becomes bimodal for droplets larger than about 15 μm. The velocity histogram for droplets between 20 and 30 μm clearly illustrates this bimodality which becomes distinct as the jet wall is approached. The larger rms fluctuations reported in Fig. 2.5 particularly for large droplets are therefore due to this bimodality. It is interesting to note here from the histogram shown in Fig. 2.11 that for droplets with the range $20 < d < 30\,\mu$m, while the overall rms fluctuations of axial velocities will be high due to the bimodality, the rms fluctuations within the individual lower velocity or higher velocity clusters are comparable to, if not lower than, those of small droplets.

The mean velocities for specific droplet size are marked by the solid lines on the scatter plots. At $r/D = 0.41$ the measured velocity data shows two separate branches on either side of the mean velocity line where majority of the data points are clustered together. It is noted that the mean axial velocity generally decreases with increasing droplet diameter. This is expected since small droplets have larger drag coefficient and thus small relaxation time and that enables them to move faster with the flow. Larger droplets with greater inertia have longer relaxation time and thus exit the nozzle with negative slip velocities.

In order to further shed light on the reasons behind this bimodal behavior, the exit conditions of the pipe are examined with high-speed imaging at 5 kHz. The beam from a dual-head, high-speed Edgewave Nd:YAG laser producing an output beam

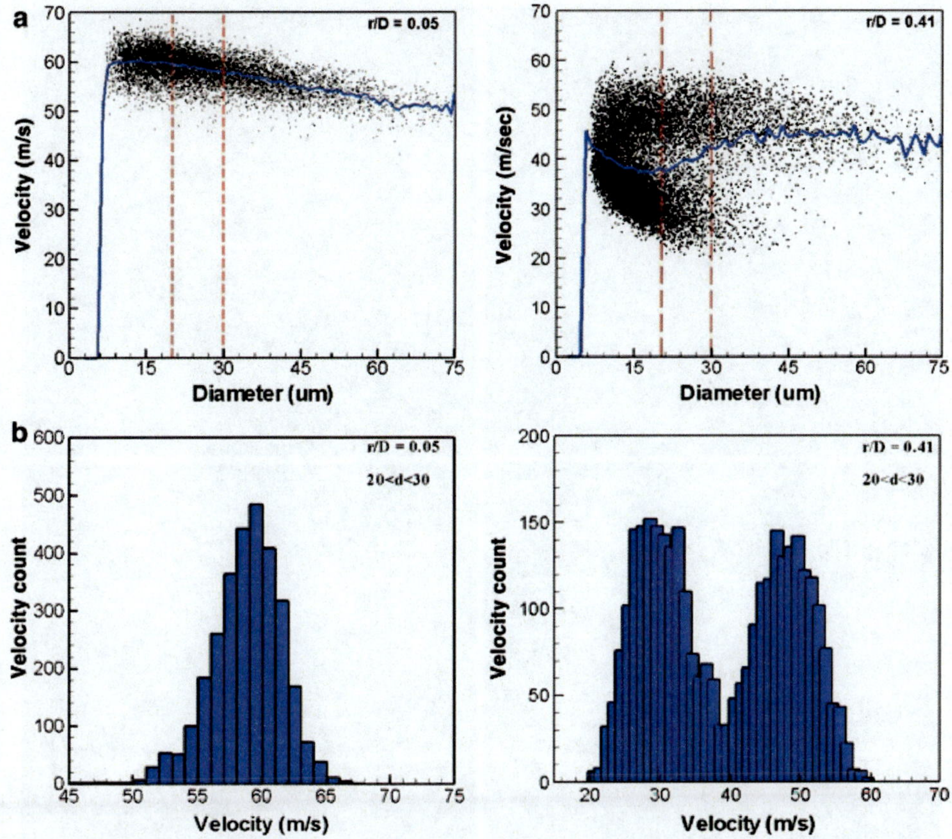

Fig. 2.12 (**a**) Real time velocity versus droplet diameter scatter plot for measurement taken close to the center of the jet at r/D = 0.057 and at the jet edge closer to the nozzle wall at r/D = 0.41. (**b**) Velocity probability function (PDF) for droplet in the size range $30 < d < 40\,\mu m$, measured at the two radial locations

(about 20 μJ per pulse at 532 nm) was focused into a sheet, 13 mm high which was directed through the centerline of the spray jet almost flush with the jet exit plane. Mie scattering from the droplets was then collected on a high speed CMOS camera (HSS6, LaVision) also operating at 5 kHz. A 532 nm interference filter with a 10 nm FWHM (CVI, F10-532.0-4-2.00) was used to ensure only the Mie scattering signal was detected. In addition, a Komura 85 mm F#1.4 lens was used to focus the image onto the camera.

Figure 2.13 shows selected images for Mie scattering collected at the exit plane of a non-reacting spray jet. The images illustrated here are not in time sequence and the vertical bars at the lower edge of the image mark the walls and the centerline of the jet. A very interesting finding revealed by these images is the occurrence of droplet 'shedding' from the internal wall of the tube marked by the horizontal arrows. Filaments or strings of droplets are emitted from the jet edge and subsequently break down into droplets. These events are characterized by a dense stream of liquid

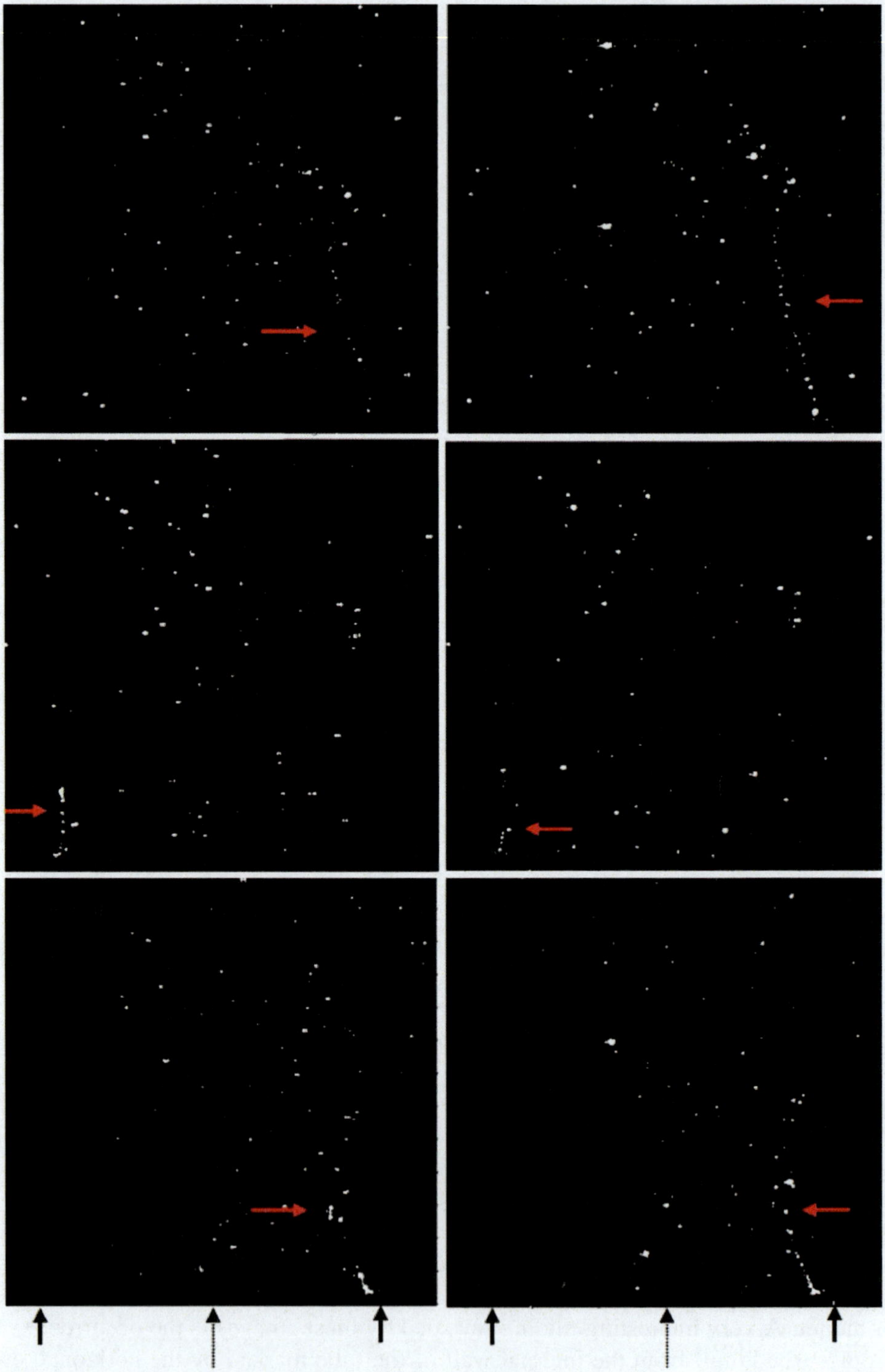

Fig. 2.13 Selected images of Mie scattering showing droplet shedding (indicated by *horizontal arrows*) from the walls of the spray jet. Tube internal diameter is 10.5 mm. Image size is 13 mm × 13 mm. *Vertical arrows* mark the edges and centre of the fuel pipe

that builds up on the inner wall and releases after the thickness of the boundary layer reaches a certain threshold. It can be seen from the images that the shedding may occur on either sides so it is not uniform around the circumference of wall's inner boundary layer. The shedding frequency of these filaments is estimated at around 250 Hz by only counting shedding from one side of the tube.

The high speed imaging results obtained at the jet exit plane confirm the following scenario for the bimodal distribution of velocities measured for large droplets near the inner walls of dilute spray jets (Fig. 2.11). A liquid boundary layer gradually develops on the inner wall of the pipe and intermittently sheds filaments of droplets from various points around the periphery of the pipe which further disintegrate into relatively large, slow droplets. Additionally, some droplets from the tube centerline move radially out towards the wall and gradually slow down. These categories would form the slower moving droplets that are detected near the wall. The faster droplets are those that are ejected from the tube without being affected by the boundary layer or the presence of the wall. Further studies are being conducted to understand the evolution of the boundary layer within the pipe and to determine means of preventing or eliminating this bimodal behavior.

6 Concluding Remarks

Comprehensive boundary conditions are presented in this chapter for non-reacting and reacting dilute sprays stabilized on a spray jet burner. The simple but representative flows and the well-defined boundaries make this burner geometry an ideal platform for development of validation of computational tools for turbulent sprays. Non-reacting as well as reacting cases for two fuels are selected to study the effects of changing the carrier velocity and the droplet loading. The Nukiyama-Tanasawa distribution is a good representative of the probability density of droplets at the jet exit plane and this remains largely unaffected with increasing droplet loading but shifts towards smaller droplets as the velocity is increased.

Velocity measurements in a non-reacting gaseous mixtures of argon and nitrogen show good agreement with velocity of small droplets less than 5 or 10 μm, hence these are seen to be adequate followers of the gaseous flow. Droplets larger than about 10 μm exit the nozzle with an increasingly negative slip velocity so that their mean axial velocity is lower than that of the gas. Close to the inner wall of the pipe, it is found that the rms fluctuations of large droplets are higher than those of small ones. This phenomenon is contrary to expectations and is referred to as the spray pipe anomaly. Instantaneous scatter plots show that in these regions the PDFs of the velocity for large droplets are bimodal and high speed imaging reveals that this bimodality and hence anomaly is due to intermittent shedding of liquid filaments from the liquid film that builds on the inner wall. Such filaments further disintegrate into large droplets which then lead to the bimodal distributions measured at the jet exit plane near the pipe wall.

Acknowledgment This work is supported by a grant from the Australian Research Council. The authors are grateful to Mr. William O'Loughlin for providing the high-speed images shown here, to Professor R.W Bilger for his insightful comments and to Dr Sten Starner for his able assistance with the experiments.

References

1. The International Workshop on Measurement and Computation of Turbulent Nonpremixed Flames. http://www.sandia.gov/TNF/abstracts.html
2. Masri, A.R., Dibble, R.W., Barlow, R.S.: The structure of turbulent nonpremixed flames revealed by Raman-Rayleigh-LIF measurements. Prog. Energy Combust. Sci. **22**, 307–362 (1996)
3. Masri, A.R., Bilger, R.W., Barlow, R.S.: Turbulent nonpremixed flames of methane near extinction: probability density functions. Combust. Flame **73**, 261–285 (1988)
4. Masri, A.R., Dibble, R.W., Barlow, R.S.: The structure of turbulent nonpremixed flames of methanol over a range of mixing rates. Combust. Flame **89**, 167–185 (1992)
5. Markides, C.N., Mastorakos, E.: An experimental study of hydrogen autoignition in a turbulent co-flow of heated air. Proc. Combust. Inst. **30**, 881–888 (2005)
6. Echekki, T., Chen, J.H.: Direct numerical simulation of autoignition in non-homogeneous hydrogen-air mixtures. Combust. Flame **134**, 169–191 (2003)
7. Gordon, R.L., Masri, A.R., Pope, S.B., Goldin, G.M.: A numerical study of auto-ignition in turbulent lifted flames issuing into a vitiated co-flow. Combust. Theory Modelling **11**, 351–376 (2007)
8. Dunn, M.J., Masri, A.R., Bilger, R.W.: A new piloted premixed jet burner to study strong finite-rate chemistry effects. Combust. Flame **151**, 46–60 (2007)
9. Dunn, M.J., Masri, A.R., Bilger, R.W., Barlow, R.S., Wang, G.H.: The compositional structure of highly turbulent piloted premixed flames issuing into a hot co-flow. Proc. Combust. Inst. **32**, 1779–1786 (2009)
10. Dally, B.B., Masri, A.R., Barlow, R.S., Fiechtner, G.J., Fletcher, D.F.: Instantaneous and mean compositional structure of bluff-body stabilised nonpremixed flames. Combust. Flame **114**, 119–148 (1998)
11. Al-Abdeli, Y.M., Masri, A.R.: Stability characteristics and flow fields of turbulent swirling jet flows. Combust. Theory Modelling **7**, 731–766 (2003)
12. Al-Abdeli, Y.M., Masri, A.R., Marquez, G.R., Starner, S.H.: Time-varying behaviour of turbulent swirling nonpremixed flames. Combust. Flame **146**, 200–214 (2006)
13. Klimenko, A.Y., Bilger, R.W.: Conditional moment closure for turbulent combustion. Prog. Energy Combust. Sci. **25**, 595–687 (1999)
14. Navarro-Martinez, S., Kronenburg, A.: LES-CMC simulations of a turbulent bluff-body flame. Proc. Combust. Inst. **31**, 1721–1728 (2007)
15. Pierce, C.D., Moin, P.: Progress-variable approach for large eddy simulation of nonpremixed turbulent combustion. J. Fluid Mech. **504**, 73–97 (2004)
16. Ihme, M., Pitsch, H.: Prediction of extinction and reignition in nonpremixed turbulent flames using a flamelet/progress variable model 2. Application in LES of Sandia flames D and E. Combust. Flame **155**, 90–107 (2008)
17. Tang, Q., Xu, J., Pope, S.B.: Probability density function calculations of local extinction and NO Production in piloted-jet turbulent methane/air flames. Proc. Combust. Inst. **28**, 133–140 (2000)
18. Lindstedt, R.P., Louloudi, S.A., Vaos, E.M.: Joint probability density function modeling of pollutant formation in piloted turbulent jet diffusion flames with comprehensive chemistry. Proc. Combust. Inst. **28**, 149–156 (2000)
19. Lozano, A., Yip, B., Hanson, R.K.: Acetone: a tracer for concentration measurements in gaseous flows by planar laser-induced fluorescence. Experiments in Fluids **13**, 369–376 (1992)

20. Linne, M.A., Paciaroni, M., Berrocal, E., Sedarsky, D.: Ballistic imaging of liquid breakup processes in dense sprays. Proc. Combust. Inst. **32**, 2147–2161 (2009)
21. Menard, T., Tanguy, S., Berlemont, A.: Coupling level set set/VOF/ghost fluid methods: validation and application to 3D simulation of the primary break-up of a liquid jet. Int. J. Multiphase Flow **33**, 510–524 (2007)
22. Lebas, R., Menard, T., Beau, P.A., Berlemont, A., Demoulin, F.X.: Numerical simulation of primary break-up and atomization: dns and modelling study. Int. J. Multiphase Flow **35**, 247–260 (2009)
23. Faeth, G.M.: Current status of droplet and liquid combustion. Prog. Energy Combust. Sci. **3**, 191–234 (1977)
24. Faeth, G.M.: Evaporation and combustion of sprays. Prog. Energy Combust. Sci. **9**, 1–76 (1983)
25. Fan, L.S., Lau, R., Zhu, C., Vuong, K., Warsito, W., Wang, X., Liu, G.: Evaporative liquid jets in gas-liquid-solid flow system. Chem. Eng. Sci. **56**, 5871–5891 (2001)
26. Lasheras, J.C., Hopfinger, E.J.: Liquid jet instability and atomization in a coaxial gas stream. Ann. Rev. Fluid Mech. **32**, 275–308 (2000)
27. Ferrand, V., Bazile, R., Borée, J., Charnay, G.: Gas-droplet turbulent velocity correlations and two-phase interaction in an axisymmetric jet laden with partly responsive droplets. Int. J. Multiphase Flow **29**, 195–217 (2003)
28. Fleckhaus, D., Hishida, K., Maeda, M.: Effect of laden solid particles on the turbulent flow structure of a round free jet. Exp. Fluids **5**, 323–333 (1987)
29. Hardalupas, A.M., Taylor, K.P., Whitelaw, J.H.: Velocity and particle flux characteristics of turbulent particle laden-jets. Proc. R. Soc. Lond. A **426**, 31–78 (1989)
30. Hetsroni, G.: Particles-turbulence interaction. Int. J. Multiphase Flow **15**, 735–746 (1989)
31. Kennedy, I.M., Moody, M.H.: Particle dispersion in a turbulent round jet. Exp. Thermal Fluid Sci. **18**, 11–26 (1998)
32. Solomon, A.S.P., Shuen, J.S., Zhang, Q.F., Faeth, G.M.: Measurements and predictions of the structure of evaporating sprays. ASME J. Heat Transfer **107**, 679–686 (1985)
33. Yuan, Z., Michaelides, E.E.: Turbulence modulation in particulate flows – a theoretical approach. Int. J. Multiphase Flow **18**, 779–785 (1992)
34. Gore, R.A., Crowe, C.T.: Effect of particle size on modulating turbulent intensity. Int. J. Multiphase Flow **15**, 279–285 (1989)
35. Chen, Y.-C., Starner, S.H., Masri, A.R.: A detailed experimental investigation of well-defined, turbulent evaporating spray jets of acetone. Int. J. Multiphase Flow. **32**, 389–412 (2006)
36. Starner, S.H., Gounder, J.D., Masri, A.R.: Effects of turbulence and carrier fluid on simple. Turbulent spray jet flames. Combust. Flame **143**, 420–432 (2005)
37. Gounder, J.D., Masri, A.R.: Turbulent piloted dilute spray flames: flow fields and droplet dynamics. Combust. Flame (submitted)
38. TSI, Phase Doppler Particle Analyzer (PDPA)/Laser Doppler Velocimeter (LDV) Operations Manual (2005)
39. TSI, Flow Sizer software, V 2.0.0.0
40. Dally, B.B.: Turbulent nonpremixed flames stabalized on a bluff-body. Ph.D. thesis, The University of Sydney, Sydney (1997)
41. Masri, A.R.: Turbulence-chemistry interactions in nonpremixed flames of hydrocarbon fuels. Ph.D. thesis, The University of Sydney, Sydney (1987)
42. Gounder, J.D.: An experimental investigation of non-reacting and reacting spray jets. Ph.D. thesis, The University of Sydney, Sydney (2009)
43. Chapman, S., Cowling, T.G.: The mathematical theory of non-uniform gases: an account of the kinetic theory of viscosity, thermal conduction, and diffusion in gases. Cambridge University Press, Cambridge (1990)
44. Nukiyama, S., Tanasawa, Y.: Trans. SME Jpn. **6**, 5–7 (1940)
45. Hinze, J.O.: Turbulence. McGraw-Hill, New York (1975)
46. Faeth, G.M.: Spray combustion phenomena. Proc. Combust. Inst. **26**, 1593–1612 (1996)

47. Starner, S.H., Chen, Y.-C., Masri, A.R.: Mass flux measurements in simple, evaporating turbulent spray jets. Proceedings of the Third Asia-Pacific Conference on Combustion, Seoul, pp. 331–334 (2001)
48. Starner, S.H., Masri, A.R.: An experimental comparison of evaporating and non-evaporating sprays in a simple turbulent jet flow. Proceedings of 9th International Conference on Liquid Atomization and Spray Systems, Sorrento, p. 1708 (2003)
49. Gounder, J.D., Masri, A.R.: Flow field and mass flux measurements near the exit plane of spray jets. 11th Triennial International Annual Conference on Liquid Atomization and Spray Systems, ICLASS 2009, CD-ROM Publication, Vail, Colorado (2009)

Chapter 3
Toward the Impact of Fuel Evaporation-Combustion Interaction on Spray Combustion in Gas Turbine Combustion Chambers. Part I: Effect of Partial Fuel Vaporization on Spray Combustion

Amsini Sadiki, W. Ahmadi, Mouldi Chrigui, and J. Janicka

Abstract This work aims at investigating the impact of the interaction between evaporation process and combustion on spray combustion characteristics in gas turbine combustion chambers. It is subdivided into two parts. The first part studies how the evaporation process affects the behavior of partially pre-vaporized spray combustion. The second part attempts to answer the question how the fuel evaporation process behaves under premixed combustion conditions.

For this purpose an Eulerian-Lagrangian RANS based procedure under a full two-way coupling was used. To describe the 3D-evaporation, two different (equilibrium and non-equilibrium) evaporation models based on the uniform temperature assumption were applied. For the combustion, the conditioned progress variable approach based on the Bray-Moss-Libby model has been adapted and used to account for both premixed and partially premixed combustion.

To assess the numerical approach and to analyze the ongoing processes in the first part of this work, a model gas turbine combustor fired by kerosene fuel was considered. It features a partially premixed flame. Among others, the influence of both parameters, the mixing air temperature and the vaporization tube length, on the degree of vaporization and subsequently on spray flame characteristics has been pointed out. An overall agreement with available experimental data was achieved especially using the non-equilibrium evaporation model. With this confidence in place, further parameter studies could be consistently performed and additional information could be extracted to gain insights relevant to understanding the effects under study.

A. Sadiki (✉), W. Ahmadi, M. Chrigui, and J. Janicka
Department of Mechanical and Processing Engineering, Institute for Energy and Powerplant Technology, Technische Universität Darmstadt, Petersenstr. 30, 64287 Darmstadt, Germany
e-mail: sadiki@ekt.tu-darmstadt.de; ahmadi@ekt.tu-darmstadt.de; mchrigui@ekt.tu-darmstadt.de; janicka@ekt.tu-darmstdt.de

A. Sadiki and J. Janicka
Center of Smart Interface, Technische Universität Darmstadt, Petersenstr. 32, 64287 Darmstadt, Germany

B. Merci et al. (eds.), *Experiments and Numerical Simulations of Diluted Spray Turbulent Combustion*, ERCOFTAC Series 17, DOI 10.1007/978-94-007-1409-0_3,

Keywords Kerosene spray combustion · Partially premixed pre-vaporization · RANS · Euler-Lagrange

Nomenclature

Symbol	Dimension	Definition
B_M	–	Spalding mass transfer number
B_T	–	Spalding heat transfer number
c_p	J/(kg.K)	Specific heat capacity by constant pressure
C_w	–	Drag coefficient
$C_{\varepsilon,3}^{k-\varepsilon}$	–	Constant model for the turbulence modulation
C_L	–	Model constant
C_R	–	Model constant
c	–	Progress variable
D	–	Fractal dimension
D_m	s^{-1}	Diffusion term in scalar transport equation
D_p	m	Droplet diameter
$\overrightarrow{F}/F_i$	N	Force
$\overrightarrow{g}/g_i$	m/s^2	Gravity acceleration vector/Cartesian components
h_v	J	Latent heat of vaporization
k	m^2/s^2	Turbulent kinetic energy
l (L)	mm	Distance coved by spray
Le	–	Lewis number
L_E	m	Turbulent length scale
$\dot{m}_{pf}$	kg/s	Droplet mass flow cross a cell
$\dot{m}_{p,v}$	kg/s	Droplet evaporation rate
N_p	–	Number of real droplets represented by one numerical parcel
p	Pa	Static pressure
PDF	–	Probability density function
Pr	–	Prandtl number
$\dot{Q}$	J/s	Heat flux rate penetrating into the droplet interior
RMS (rms)		Root mean square
Re	–	Reynolds number
Re_p	–	Droplet Reynolds number
Sc	–	Schmidt number
Sh	–	Sherwood number
SMD	mm	Droplet Sauter mean diameter
St	–	Stokes number
$S_{\psi,p,s}$	ψ dependent	Source term of the variable ψ for a non-evaporating droplet
$S_{\psi,p,v}$	ψ dependent	Source term of the variable ψ due to evaporation
s_r^o	m/s	Laminar burning velocity
q_x	mm^3/m	Droplet axial volume flux
T, T_G, T_d	K	Temperature, gas phase temperature, droplet temperature
T_L	s	Lagrangian integral time scale
t	s	time

(continued)

(continued)

Symbol	Dimension	Definition
$\overrightarrow{u}/(u,v,w)$	m/s	Velocity vector/Cartesian components
u', v', w'	m/s	Gas phase velocity fluctuation components
$\overrightarrow{u}_p/(u_p, v_p, w_p)$	m/s	Droplet velocity vector/Cartesian components
u'_p, v'_p, w'_p	m/s	Dispersed phase velocity fluctuation components
$u'^F_{i,F}$	m/s	Fluid element velocity fluctuation along the stream line
$u'^F_{i,p}$	m/s	Fluid element velocity fluctuation at the droplet location
$V_{i,j,k}$	m^3	Volume of the cell ijk
x_j	m	Coordinate components
Y	–	Vapor mass fraction
$\tilde{z}$	–	Filtered mixture fraction
$\widetilde{z''^2}$	–	Filtered mixture fraction variance
α_p	–	Droplet volume fraction
ε	m^2/s^3	Turbulent kinetic energy dissipation rate
Δt	s	Displacement in time
Δr	m	Displacement in space
μ/μ_t	kg/(m.s)	Dynamic molecular/turbulent viscosity
ν/ν_t	m^2/s	Kinematic molecular/turbulent viscosity
$\upsilon_{s,eq}$	–	Molar mass fraction
σ		Standard deviation
ρ	kg/m^3	Density of mass
$\underset{\sim}{\phi}$	–	Equivalence ratio
$\dot{w}_c$	$kg/(m^3.s)$	Reaction rate source term

1 Introduction

Spray combustion is used extensively in many technical combustion engines, such as gas turbine, automotive engines, industrial furnaces, and liquid-fueled rocket engines. Thereby, the liquid fuel is supplied with varying physical and chemical properties to form a combustible mixture of fuel vapor and air after undergoing various processes, like atomization, primary and secondary break up, droplet dispersion along its impact on turbulence, vaporization and mixing [[1–11], and references therein].

The understanding of the fuel-air mixture preparation process and its interaction with combustion through the evaporation process is milestone for the improvement of combustion efficiency and pollutant emission reduction. Indeed, poor mixing of the fuel and oxidant streams leads to mixture inhomogeneities that affect combustion efficiency and result in enhanced pollutants formation. Especially, combustion processes with air excess exhibit a high potential for the reduction of the NOx-emissions from spray flames in gas turbines. The concept of lean premixing has been successfully considered for utility gas turbines operating with both gaseous and liquid fuels, respectively. In [2] it is reported that the NOx-emissions are typically reduced to 25 and 50 ppm, 15% O_2, if full pre-vaporization and premixed can be achieved. Since they are prone to autoignition and flashback, fully-premixed systems are unlikely to appear in the near future. A promising method may consist in partial premixing of sprays with extremely low mixing times. However, the promise of this method depends obviously on the link to the degree of vaporization to be achieved for a given NOx-limit.

As pointed out in [1–7, 9, 12–35] a considerable amount of works have been done including diverse parameter studies. However, there are relatively few experimental and numerical results devoted to the effects of degree of partial vaporization on spray combustion and pollutant emission. It is worth mentioning that an accurate capture of the turbulence and its modulation characteristics is determinant for an accurate prediction of evaporation, that, in turn, affects significantly the spray combustion along with the pollutant emissions [1–5]. In order to provide early detailed information at moderate costs of all these processes, CFD has become an important tool besides experiments [6–32,36–58]. With reasonably accurate achievements in simulating single phase combustion in complex configurations [43–45, 47–53], CFD tools have shown to lack in reliability regarding the prediction of spray combustion properties [12–18, 33]. Especially the question how (partial) vaporization may influence spray flame properties and pollutant emission of spray flames is not yet well addressed.

A comprehensive insight can be directly attempted by a direct numerical simulation (DNS). However, a DNS approach for capturing the turbulence under conditions of practical applications is nowadays economically unviable and advanced [10, 11, 35]. Physically consistent and fast sub-models are still needed, and the prediction capability of these models has to be carefully evaluated.

There is today enough evidence that Large Eddy Simulation (LES) is able to well capture intrinsically time- and space dependent phenomena [12–14, 42–45]. However, the compromise to be found between the grid size of complex geometries and the computational time does not allow nowadays to perform LES for daily engineering tasks, once focused on liquid fuel sprays [12–14]. Thus, Reynolds-Averaged based Numerical Simulations (here RANS) appear to be for a long time the compromise technique [8, 15–31]. It needs to be further addressed in order to improve its predictive capability. This effort is pursued in this work.

In the first part of this contribution, spray combustion in a model gas turbine combustor exhibiting a partially premixed combustion is under study. It is fired by kerosene fuel [2], and so designed that atomization and break up processes [30, 31] are negligible.

With regard to kerosene sprays, Becker and Hassa [30], among others, studied kerosene droplet SMD distributions in airblast sprays under elevated pressures and gas turbine conditions. A large amount of experimental research is being conducted to strive low NOx production by designing devices that permit rapid spray phase transition, mixing and combustion. Baessler et al. [2] studied the NOx emission of premixed partially vaporized kerosene spray flame at atmospheric conditions and found out that dealing with lean conditions a reduction of NOx requires that a prevaporization of the kerosene spray should pass over 50% upstream of the flame. They realized that increasing the prevaporization zone length reduces the emissions. Rokke and Wilson [4] investigated venture LPP on mixing, atomization and evaporation behavior as well as emission, while Schäfer et al. [3] concentrated on the flashback phenomena. Nomura et al. [5] studied experimentally a partially prevaporized spray burner with monodispersed ethanol droplets to investigate the interaction between fuel droplets and a flame. They investigated the effect of mean droplet diameter, and the entry length of droplets into a flame on the laminar burning velocity of partially pre-vaporized sprays.

Real liquid fuels mostly used, whether they are petroleum-derived or biomass-derived, are composed by many compounds with a wide range of properties. Developing evaporation models and reaction mechanism under consideration of all the compounds is a huge task that is still challenging (see e.g. [12, 54]). In order to simulate multi-component (MC) fuels many investigators rather prefer to use a variety of heavy (large molar weight) single component (SC) or mixtures known as surrogate mixtures that adequately represent the desired physical and/or chemical characteristics. As pointed out in [59] the accuracy of the simulated results obtained from a chemical kinetic model for a surrogate depends in general on the comprehensiveness of the sub-mechanisms of individual fuel components considered in the model. According to [9, 22–24, 59] we assume such surrogate and remain in this work focus on SC evaporation models.

A classical evaporation model (rapid mixing model) was first derived by Godsave [26] and Spalding [27]. This model, also referred to as the "D_p^2-law", was originally derived assuming a constant droplet temperature fixed at the wet bulb condition, and included the quasi-steady assumption for the gas phase. Since its introduction, the importance of transient droplet heating has been recognized [28], and the evaporation rate is now generally coupled with a time dependent energy equation typically with assumed infinite thermal conductivity of the liquid (e.g. Aggarawal et al. [25]). In this form, the classical model is generally referred to as either the infinite conductivity model, or the rapid mixing model. Aggarawal et al. evaluated several evaporation models for stochastic spray simulations, including both constant droplet temperature and transient heating versions of the classical model (rapid mixing model). They recommended the use of a spherically symmetric formulation with finite liquid conductivity when the droplet is stationary and an axisymmetric internal circulation model when the droplet Reynolds number is substantially larger than unity. Unfortunately, neither of these forms is appropriate when many droplets are involved due to the extreme computational expense of resolving both the droplet interior and exterior boundary layers in either one or two dimensions.

In term of complexity, a variety of models exist as recently reviewed by Sazhin [9] and Oefelein and Aggarwal [29]. We restrict ourselves to the model by Abramzon and Sirignano [24] who revised the infinite conductivity model to incorporate the effects of Stefan flow on heat and mass transfer. In this model, heat transfer is modified through the use of modified forms for the Nusselt, Sherwood and transfer numbers. The model derived by using the Clausius-Clapeyron equilibrium equation is mostly favored for droplet evaporation simulation due to its compact formulation and direct analogy to the energy equation. It will be labeled later on as "equilibrium evaporation model".

When dealing with evaporation rates that are large or with a prompt changing of the temperature on the liquid/gas interface, non-equilibrium processes cannot be neglected [19–23]. The situation for non-equilibrium conditions due to large evaporation rate is, in fact, related to the droplet size, since the more the droplet gets smaller, the higher is the phase transition. This condition, which is being stronger at the end of the droplet life time, is perfectly fulfilled for the application of a non-equilibrium evaporation model [9, 19–23]. The second condition for non-equilibrium evaporation processes, which is present within the test case under investigation, is a prompt changing on the temperature of the liquid gas interface. This situation is often detectable when droplets leave the pre-vaporization zone and approach to the flame front, where they are faced with a very high temperature gradient. Thereby the droplet properties do not have enough time to relax toward their thermodynamic equilibrium [4, 5, 19].

Bellan and Sommerfeld [20] first introduced the non-equilibrium Langmuir-Knudsen evaporation law for use in droplet combustion models and found non-equilibrium effects to be important for droplet sizes found in practical spray configurations. For very small droplet sizes that occur during condensation and nucleation processes, Jackson and Davidson [21] incorporate the non-equilibrium Herz-Knudsen law (applicable to the free molecule regime) in their Eulerian-Eulerian gas liquid flow model. Miller et al. [19] evaluated the predictability capacity of equilibrium and non-equilibrium evaporation models for many droplet gas-liquid flow simulations. They found out that the thermodynamic non-equilibrium effects are important for initial droplet diameters $D_p < 50\,\mu m$. Sadiki et al. [23] investigated the effects of turbulence on the turbulent droplet dispersion, vaporization and mixing for non-reacting sprays and coupling of the turbulence modulation on the turbulent spray combustion. In particular they could show that for small droplets the non-equilibrium effects are not negligible in the dynamic vaporization process.

Dealing with dilute sprays, the presence of group combustion as described by Chiu and Liu [32] is not considered. A single droplet temperature sink effect is also not accounted for. We rather rely on the fact that all droplets are completely evaporated. In this case, several approaches for chemistry and combustion modeling already published in the literature are useful. Among them, the so-called conditioned progress variable approach (CPVA) based on the Bray-Moss-Libby (BML) model will be adapted and used in this work to account for both premixed and partially premixed combustion [51, 52]. A comprehensive overview of various combustion existing models within RANS and LES context can be found in [44] and [45] and therein quoted references.

The present contribution aims at investigating the impact of evaporation process on the flame characteristics under partially premixed combustion conditions. For that purpose, a RANS-based approach including a spray module developed by Sadiki et al. [23] in the framework of the collaborative research centre SFB568/TU-Darmstadt is applied. It includes advanced models for turbulence, turbulence modulation, dispersion, evaporation and combustion. To assess the prediction ability of the model combinations employed a model gas turbine combustion chamber operating under conditions similar to those found in industrial applications is considered.

The contribution is organized as follows: Sect. 2 describes the mathematical models and the numerical procedures. Especially the two different models used for the description of the evaporation and the combustion are outlined. The configuration under study along with inflow and boundary conditions are presented in Sect. 3. Detailed comparisons of numerical results with experimental data and various parameter studies are provided in Sect. 4. Section 5 is devoted to conclusions.

2 Mathematical Models and Numerical Procedures

Among various approaches used to describe numerically turbulent multiphase flows [6, 7, 58], the approach adopted in this work is based on an Eulerian-Lagrangian method.

2.1 Governing Equations for Turbulent Gaseous Flow

The carrier phase is considered as continuum phase and described using the Reynolds averaging method. For this purpose, the governing transport equations have been solved for mass, momentum, mixture fraction and its variance, reaction progress variable and turbulent quantities. The volume variation of the carrier phase as consequence of the presence of liquid droplets is neglected. The non-stationary, general form of the transport equation applied emerges as [23]:

$$\frac{\partial(\bar{\rho}\,\phi)}{\partial t} + \frac{\partial(\bar{\rho}\bar{u}\,\phi)}{\partial x} + \frac{\partial(\bar{\rho}\bar{v}\,\phi)}{\partial y} + \frac{\partial(\bar{\rho}\bar{w}\,\phi)}{\partial z} - \frac{\partial}{\partial x}\left(\Gamma\frac{\partial\phi}{\partial x}\right) - \frac{\partial}{\partial y}\left(\Gamma\frac{\partial\phi}{\partial y}\right)$$
$$-\frac{\partial}{\partial z}\left(\Gamma\frac{\partial\phi}{\partial z}\right) = S_\phi + \overline{S}_{\phi,p} + \overline{S}_{\phi,p,v} + \overline{S_c} \tag{3.1}$$

in which ϕ may represent the mean value of mass, velocity, mixture fraction, mixture fraction variance, turbulent kinetic energy, turbulent dissipation rate, and reaction progress variable, respectively. Γ represents an effective diffusion coefficient and S_ϕ the well-known turbulence source term in single phase flow cases [7, 8, 38, 42, 43]. The additional source terms characterize the direct interaction of mass, momentum, transported mixing and combustion scalar quantities and turbulent quantities between the two phases and therefore account for the coupling between the fluid turbulence and the evaporating droplets moving in the turbulent flow. The first term, $\overline{S}_{\phi,p}$, expresses especially the classical two-way coupling in absence of evaporation and combustion; the term, $\overline{S}_{\phi,p,v}$, takes into consideration the transfer caused by the phase transition processes, like evaporation described below. Details about these terms and their relationship can be found in [8, 13–19, 23], among others. They are listed in Table 3.1 for self-consistency purpose of the present contribution. The last term $\overline{S}_c$ is the chemical source term.

Efforts to better capture streamline curvature effects and swirled flows phenomena have been achieved in the last years by means of second order models or by their corresponding algebraic formulations, besides advanced non-linear turbulence and mixing models [42]. We apply here the RNG (ReNormalization Group) k-Epsilon

Table 3.1 Source terms due to the presence of evaporating droplets

ϕ	$\overline{S}_{\phi,p}$	$\overline{S}_{\phi,p,v}$
1	0	$-\sum_p \frac{\dot{m}_{p,v} N_p}{V_{i,j,k}}$
$\bar{u}$	$-\sum_p \frac{\dot{m}_p N_p}{V_{i,j,k}} \left[\left(u_p^{t_n+\Delta t} - u_p^{t_n} \right) - g_x \Delta t \right]$	$\sum_p \frac{\dot{m}_{p,v} N_p}{V_{i,j,k}} u_p$
$\bar{v}$	$-\sum_p \frac{\dot{m}_p N_p}{V_{i,j,k}} \left[\left(v_p^{t_n+\Delta t} - v_p^{t_n} \right) - g_y \Delta t \right]$	$\sum_p \frac{\dot{m}_{p,v} N_p}{V_{i,j,k}} v_p$
$\bar{w}$	$-\sum_p \frac{\dot{m}_p N_p}{V_{i,j,k}} \left[\left(w_p^{t_n+\Delta t} - w_p^{t_n} \right) - g_z \Delta t \right]$	$\sum_p \frac{\dot{m}_{p,v} N_p}{V_{i,j,k}} w_p$
$\tilde{z}$	0	$\sum_p \frac{\dot{m}_{p,v} N_p}{V_{i,j,k}}$
$\widetilde{z''^2}$	0	$\left(-\sum_p \frac{\dot{m}_{p,v} N_p}{V_{i,j,k}} \right) \frac{\widetilde{z''^2}(1-2\tilde{z})}{\tilde{z}}$ [15, 33]
k	$\overline{u_i S_{u_i,p}} - \overline{u_i}\,\overline{s_{u_i,p}}$	$\overline{u_i S_{u_i,m}} - \overline{u s_{u_i,m}} + \frac{1}{2} \left(\overline{u_i^2}\,\overline{s_{p,m}} - \overline{u_i^2 s_{p,m}} \right)$
ε	$c_{\varepsilon,3}^{k-\varepsilon} \frac{\varepsilon}{k} S_{k,p}$; $c_{\varepsilon,3}^{RS} \frac{\varepsilon}{k} S_{k,p}^{RS}$	$c_{\varepsilon,3}^{k-\varepsilon} \frac{\varepsilon}{k} S_{k,p,v}$; $c_{\varepsilon,3}^{RS} \frac{\varepsilon}{k} S_{k,p,v}^{RS}$

model [57] modified for two phase flow description by including additional source terms mentioned above [16]. For simplicity, the turbulent scalar flux vector in transport equations of mixing and combustion scalar quantities has been postulated by means of a gradient assumption.

2.2 *Two-Way Coupling and Turbulence Modulation*

The key physics in reacting dispersed multiphase flows are in the coupling between the phases through the mass, momentum and transported mixing and combustion scalar quantities. Furthermore, extra mechanisms of turbulence modulation including turbulence production, distortion, and dissipation, that might influence strongly the evaporation process, may become important, and the turbulent stresses can even be either reduced or increased [6–8, 36–38]. This is the case in configurations with higher volume fraction or when $D_p \geq \eta$ (η: Kolmogorov length scale).

According to the classification criteria of interphase coupling by Elgobashi or to Crowe (see in [6, 7, 36]), the flow systems under study belong to dilute two-phase flows in which the droplet volume fraction is of the order of 10^{-4}. Therefore a full two-way phase coupling should be considered.

In polydisperse sprays several different mechanisms can cause turbulence modulation, and sometimes multiple mechanisms act simultaneously, so that an overbalancing of droplet-induced turbulence attenuation and production should be expected. As pointed out in [6–8, 36–38] this complex phenomena cannot be well captured by the state-of-the art approaches.

Because both droplet-induced attenuation and production of continuous phase turbulence due to interphase couplings or/and to interphase transport phenomena are thermodynamic processes, we follow [39] and apply in this work a thermodynamically consistent modulation model [8, 23]. This model has been designed in [7, 8, 23] based on the exploitation of the entropy inequality expressing the second law of thermodynamics for multiphase flows. The particle/droplet source term for the turbulent kinetic energy is then given in this model by [8, 23]:

$$S_{k,p} = \beta \left(\overline{u_{pi} S_{u_i,p}} - \overline{u_i}\ \overline{S_{u_i,p}} \right) + \left(\overline{u_i S_{u_i,p}} - \overline{u_i}\ \overline{S_{u_i,p}} \right) \tag{3.2a}$$

where

$$\beta = \alpha' + \frac{(1-\alpha') \left(\overline{u_{pi} S_{u_i,p}} - \overline{u_{pi}}\ \overline{S_{u_i,p}} \right)}{\left(\overline{u_{pi} S_{u_i,p}} - \overline{u_i}\ \overline{S_{u_i,p}} \right)} \tag{3.2b}$$

is a model parameter. Such a consistent approach is proven to improve the prediction of mass and heat transport processes involving evaporation and combustion, which in turn affect the turbulence. The second term in Eq. 3.2a represents the usual dissipative standard contribution as given in Table 3.1, while the first term accounts for the production of the turbulent kinetic energy. So, this thermodynamically consistent model captures well both the enhancement and the diminution of the turbulence of the gas phase due to the presence of both big and small droplets in polydispersed sprays. The parameter α' in Eq. 3.2b depends on the droplets properties, as shown in [8]. The relevance of this model has already been demonstrated in the frame of RANS [8, 23] and LES [40]. Note that in the model by Crowe reformulated by Lain and Sommerfeld [38] in a Lagrangian form, the second term in Eq. 3.2a vanishes while $\beta = 1$.

2.3 *Droplet Description, Dispersion and Evaporation*

As pointed out above, a Lagrangian approach is employed to compute properties of droplets moving in the turbulent flow. All numerical parcels are tracked by solving their equations of motion. One parcel represents a packet of droplets having identical characteristics.

2.3.1 Droplet Description

Following the Basset-Boussinesq-Oseen equations, the droplet position $x_{p,i}$ and velocity $u_{p,i}$ are described by (see e.g. [6, 8, 12–19, 58]):

$$\frac{dx_{p,i}}{dt} = u_{p,i} \text{ and } \frac{du_{p,i}}{dt} = \frac{3}{4} \frac{C_W}{D_p} \frac{\rho}{\rho_p} \left| \vec{u} - \vec{u_p} \right| (u_i - u_{p,i}) + \frac{(\rho_p - \rho)}{\rho_p} g_i = \frac{1}{m_p} \sum_i F_i \tag{3.3}$$

where the other forces, like the Basset history term, the added mass effect and the unsteady drag effect are negligibly small for large density ratios. The drag coefficient used within this work is determined for a spherical, not deformable, particle by:

$$C_W = \frac{24}{\mathrm{Re}_p}\left(1 + \frac{1}{6}\mathrm{Re}_p^{2/3}\right) \qquad \mathrm{Re}_p \leq 1{,}000$$
$$C_W = 0.44 \qquad \mathrm{Re}_p \geq 1{,}000 \tag{3.4}$$

where

$$\mathrm{Re}_p = \frac{\rho D_p \left|u - u_p\right|}{\mu}$$

denotes the particle Reynolds number.

2.3.2 Turbulent Droplet Dispersion

In the Lagrangian droplet tracking approach the instantaneous fluid velocity is unknown. It influences the droplet trajectories that are especially important when evaporation takes place, when temperature and chemical composition of the droplets depend on their history. In order to quantify the instantaneous fluid velocity seen by the droplets as it appears in Eq. 3.3 and its effect on the droplet distribution one should model the Root Mean Square (RMS) values of the fluid parcel velocity at the droplet location. This is done by dispersion models. Various models have been proposed in the literature. Minier [56] suggested a Langevin equation model for the instantaneous fluid velocity working on particles. He extended the "Simplified Langevin Model" for fluid particles (see in [55]) to inertial particles. Simple models, such as the "eddy-life time" approach (see in [18, 58]) and the "random walk" models, also called Lagrangian stochastic models [46], have shown good results for particles in grid turbulence [55]. However, the prediction of the RMS of particles velocity fluctuations failed and showed that the eddy-life time approach is insufficient in anisotropic turbulent flows.

Here we apply the Lagrangian stochastic approach using the fluid turbulent variables following [8, 23, 41, 55]. The discretized equations for a Markov-chain random-walk of fluid particle velocity fluctuations for non-homogeneous turbulence are given as:

$$u_i'^F(t_{n+1}) = u_i'^F(t_n) \cdot R_{p,i}(\Delta t, \Delta r) + \sigma_i \sqrt{1 - R_{p,i}^2(\Delta t, \Delta r)} \cdot \chi_i(t_n)$$
$$+(1 - R_{p,i}(\Delta t, \Delta r)) T_{L,i} \frac{\partial u_i' u_i'}{\partial x_i}. \tag{3.5}$$

whose solution for given initial conditions represents a Markov sequence of successive fluid particle velocity at time interval of Δt. Hence, this model can be seen as a Langevin model for an Orstein-Uhlenbeck process written in finite difference form.

Hereafter it will be referred as Markov-sequence dispersion model. In Eq. 3.5 the last term on the right-hand side was introduced to extend the well-established homogeneous turbulence formulations for flow anisotropy and then to avoid non-physical behavior (i.e. the accumulation of particles in regions of low velocity variance). In this equation, $\chi_i(t_n)$ denotes a Gaussian random variable, t_{n+1} and t_n are the consecutive time steps. Δr represents the spatial displacements during the time interval Δt and

$$R_{p,i}(\Delta t, \Delta r) = R_{L,i}(\Delta t) \cdot R_{E,i}(\Delta r) \tag{3.6}$$

is a typical correlation used to compute the fluctuation of the fluid element ($u_i'^F$) at the droplet location. Thereby, the evolution of the fluid element velocity fluctuation along the stream line is determined using the Lagrangian correlation factor $R_{L,i}(\Delta t)$, while the fluid element velocity fluctuation located at the droplet position is correlated with the fluid element location using the Eulerian correlation factor $R_{E,i}(\Delta r)$. They are expressed by [8, 16, 23, 55]

$$R_{L,i}\,(\Delta t) = \exp\left(-\frac{\Delta t}{T_{L,i}}\right) \quad \text{and} \quad R_{E,i}(\Delta r) = (f(\Delta r) - g(\Delta r))\frac{\Delta r_i \Delta r_j}{\Delta r^2} + g(\Delta r)\delta_{ij}, \tag{3.7}$$

respectively. The quantity $T_{L,i}$ represents the Lagrangian integral time scale calculated for a k-Epsilon like-turbulence model as $T_{L,i} = c_T k/\varepsilon = c_T \sigma^2/\varepsilon$ with $c_T = 0.3$ and $\sigma^2 = k$. The longitudinal and transversal correlation functions $f(\Delta r)$ and $g(\Delta r)$ in (3.7) are computed using the following expressions:

$$f(\Delta r) = \exp\left(-\frac{\Delta r}{L_E}\right), \quad g(\Delta r) = \left(1 - \frac{\Delta r}{2L_E}\right)\exp\left(-\frac{\Delta r}{L_E}\right), \tag{3.8}$$

where $L_E = c_L T_L \sigma$ (with $c_L = 0.3$) represents a simplified turbulent length scale.

2.3.3 Evaporation

To account for the 3D-evaporation of droplets, let us first observe that a small region of the droplet near the droplet surface will heat up quickly while the central core of the droplet remains "cold" as a single droplet enters a hot environment [9, 18, 24–29]. As the droplet penetrates further into the hot ambient gas the heat conduction to its interior increases. The temperature will eventually become nearly uniform within the droplet before the end of its lifetime. As outlined in [28], it is not trivial to solve such a transient phenomena. Among the various models that have been proposed (see in [9, 19, 25]), the so-called Uniform Temperature (UT) model by Abramson and Sirignano [24] has been adopted in this work for its accuracy and economical computing costs.

The UT model describes the evolution of the droplet's temperature and diameter, i.e. evaporation rate and energy flux through the liquid/gas interface. This model is based on the film thickness theory. All the main assumptions underlaying this theory are valid in the investigated configurations: (a) Droplets are assumed spherical; (b) secondary atomization and coalescence of droplets are neglected, i.e. simple elastic collisions between droplets and wall are assumed without any kind of film formation. In fact the Weber number near the nozzle (x = 1 mm), which is used as an indicator for the break-up phenomenon, varies between 0.01 and 0.2 in the configuration while the critical value is about 100 orders of magnitude larger, i.e. $We_{cr} = 12.07$; (c) the influence of the surface tension is neglected, i.e. uniform pressure around the droplet is assumed; (d) uniform physical properties of the surrounding fluid and liquid-vapor thermal equilibrium on the droplet surface are assumed; (e) the ambient air is not soluble in the droplet fluid; (f) chemical reactions and radiation in the droplet are not considered; (g) single component model is considered (the infinite conductivity model).

Droplet Mass Transfer

The equation governing the mass transport of the droplet is given by

$$\frac{dm_p}{dt} = -\overset{\bullet}{m}_p \text{ where } m_p = (4/3)\pi r_p^3 \rho_p \tag{3.9}$$

The quantity $\overset{\bullet}{m}_p$ is the net mass transfer rate for a droplet in a convective flow field. According to the film theory it is given by [12–18]

$$\overset{\bullet}{m}_p = 2\pi r_p \overline{\rho_m}\,\overline{D_m} Sh^* \ln(1 + B_M) \tag{3.10}$$

where $B_M = \frac{(y_s - y_\infty)}{(1-y_s)}$ and $y_s = \frac{\chi_s}{\chi_s - (1-\chi_s)\theta}$ are the mass transfer Spalding number and the surface mass fraction, respectively. Thereby θ is the ratio of molecular weights. y_∞ is the mass fraction in the gas phase. $\overline{\rho_m}$ and $\overline{D_m}$ are the averaged values of the mixture density and binary diffusion coefficient throughout the film. In this study, equilibrium and non-equilibrium considerations during evaporation are taken into account. For a detailed theoretical analysis and a comparative evaluation of different evaporation models, the reader may refer to [9, 19] and therein cited references. For the equilibrium models usually used, the molar mass fraction, χ_s, is related to the saturation pressure through the Clausius-Clapeyron equation by assuming local equilibrium between the droplet and the ambient gas. In order to incorporate the effects of Stefan flow on heat and mass transfer, Abramzon and Sirignano [24] introduced modified Nusselt Nu^* and Sherwood Sh^* numbers defined below. In the case of non-equilibrium evaporation model [19, 23], the molar mass fraction χ_s is determined by the following relation:

$$\chi_{s,neq} = \chi_{s,eq} - \left(\frac{L_K}{d/2}\right)\beta_L \quad \text{where} \quad \beta_L = -\left(\frac{3\,\mathrm{Pr}_G\,\tau_d}{2}\right)\frac{\dot{m}_p}{m} \tag{3.11}$$

represents the half of the blowing Peclet number. $\mathrm{Pr_G}$ is the Prandtl number for the gas phase, $\mathrm{L_K}$ represents the Knudsen length and $\tau_d = \frac{\rho_d D_d^2}{18\mu}$ the particle relaxation time.

Droplet Temperature

To account for both equilibrium and non-equilibrium effects the Lagrangian equation describing the transient temperature of a single droplet is modified and given by (see in [8, 10, 19]):

$$\frac{dT_d}{dt} = \frac{F^* Nu}{3\,\mathrm{Pr}_G}\left(\frac{\Pi}{\tau_d}\right)(T_G - T_d) + \left(\frac{h_v}{c_{p,l}}\right)\frac{\dot{m}_p}{m} \tag{3.12}$$

where F^* represents a correction for the effect of droplet evaporation on the droplet heating.

$$F^* = -\left(\frac{3\,\mathrm{Pr}_G\,\tau_d}{Nu}\right)\frac{\dot{m}_p}{m}\Big/ B'_T \text{ for equilibrium case}$$
$$F^* = -\left(\frac{3\,\mathrm{Pr}_G\,\tau_d}{2}\right)\frac{\dot{m}_p}{m}\Big/\left(e^{-\left(\frac{3\,\mathrm{Pr}_G\,\tau_d}{2}\right)\frac{\dot{m}_p}{m}} - 1\right) \text{ for non} - \text{eqilibrium case} \tag{3.13}$$

Thereby B'_T is the heat transfer Spalding number. It is related to $\mathrm{B_M}$ and calculated iteratively with the help of the following system equations [8, 19]:

$$B'_T = (1 + B_{M,eq})^{\varphi - 1} \qquad \varphi = \frac{c_{p,v}}{\overline{c}_p}\frac{Sh^*}{Nu^*}\frac{1}{\overline{\overline{Le}}}$$
$$Nu^* = 2 + \frac{Nu_0 - 2}{F_T} \qquad F_T = F_T(B_T) = (1 + B_T)^{0,7}\ln\frac{(1 + B'_T)}{B'_T}$$
$$Sh^* = 2 + \frac{Sh_0 - 2}{F_M} \qquad F_M = F_M(B_{M,eq}) = (1 + B_{M,eq})^{0,7}\ln\frac{(1 + B'_{M,eq})}{B'_{M,eq}} \tag{3.14}$$

In these relations, Sh_0 represents the Sherwood number in negligible evaporation (i.e. the Stefan flow is not accounted for):

$$Sh_0 = 1 + (1 + \mathrm{Re}_p Sc_p)^{1/3} f(\mathrm{Re}_p), \tag{3.15}$$

and

$$Nu_0 = 1 + (1 + \mathrm{Re}_p Pr)^{1/3} f(\mathrm{Re}_p) \tag{3.16}$$

expresses the Nusselt number for a spherical droplet in negligible evaporation. Note that Sc_p is the Schmidt number, and $f(\mathrm{Re}_p)$ is an empirical function defined as

$f(\mathrm{Re}_p) = \mathrm{Re}_p^{0.077}$. The quantity Π represents the ratio of the gas heat capacity $c_{p,v}$ to that of the liquid phase $c_{p,l}$, Le is the Lewis Number and h_v the latent heat of evaporation.

An important issue to droplet evaporation modeling is the specification of the physical properties used in the above equations. It was shown in the literature that evaporation-rate predictions are sensitive to the choice of property values of gas and vapor. The general approach is to define a reference temperature and a vapor mass fraction, which are used to evaluate both the gas and vapor properties [9]. Given the vapor and gas properties evaluated at the reference temperature, the physical properties for the mixture can be calculated using the reference mass fraction weighted averaging procedure. As pointed out by Luo et al. [10], this procedure is generally needed at each time step and can significantly increase computational cost. For the determination of thermodynamic properties of the air-vapor mixture while modeling the evaporation process, the Simpson or Sparrow and Gregg "1/3 rule" was applied [8, 9, 15–24].

2.4 Combustion Modeling

The spray flame in complex configurations is quite complicated. Generally, there are not only isolated diffusion flames and premixed flames, but also their composite structures [10, 11, 13, 33, 35]. Thereby combustion and droplet evaporation should interact for larger values of characteristic evaporation time [23] as pointed out in [35]. Although consistent results with experimental data have been reported in the literature, using models of gaseous combustion to describe spray flames leads to some relevant modeling issues. In particular, the direct influence of droplet evaporation source on spray combustion, the role of the group combustion and of a flame index characterization have to be well accounted for. These issues are not addressed in the flamelet based combustion modeling used in this study. We restrict ourselves to cases in which all droplets are supposed to have evaporated before combustion. As the reliability of spray combustion models lies essentially on an accurate description of the fuel-air mixture preparation, focus is put here on how to better retrieve the behavior of mixture fraction variance when spray evaporation is present. Therefore, we consider explicitly the influence of droplet evaporation source term primarily in the equations of the mixture fraction and mixture fraction variance.

In the configuration under investigation, one realizes further two main features, namely inhomogeneity of the equivalence ratio and of the flame propagation velocity. Since the fuel and the oxidizer necessitate certain time for the mixing, the mixture forms a spatial variation of the equivalence ratio of the fresh gas featuring a partially premixed combustion.

For the description of the resulting flame, the so-called conditioned progress variable approach (CPVA) [51] based on the Bray-Moss-Libby (BML) model is adopted. Let us outline the main aspect of the modeling procedure. Following [48], the thermo-chemistry is implemented by means of a single progress variable c that is defined as a dimensionless temperature

$$c = \frac{T - T_u}{T_b - T_u} \tag{3.17}$$

where indices u and b stand for the unburned and burnt part of the flame, respectively. Using the assumption made in [48–52] for given expansion ratio

$$\tau = \frac{T_b - T_u}{T_u} \tag{3.18}$$

the algebraic relation for the mean density can be derived as

$$\bar{\rho} = \frac{\rho_u}{1 + \tau \tilde{c}}. \tag{3.19}$$

A Favre-averaged transport equation for the single progress variable is then given by

$$\frac{\partial(\bar{\rho}\tilde{c})}{\partial t} + \frac{\partial(\bar{\rho}\tilde{u}_j\tilde{c})}{\partial x_j} = \frac{\partial}{\partial x_j}\left(-\bar{\rho}\widetilde{c''u''_j}\right) + \widetilde{\dot{w}_c} + \widetilde{\dot{w}_{c,v}} \tag{3.20}$$

Note that additional terms, like cross-scalar dissipation rate and mixture fraction dissipation rate related contributions, that emerge in a general expression of the progress variable equation as derived in [35] are neglected in (3.20). We recall that the scalar turbulent flux in (3.20) is simply closed according to the gradient diffusion hypothesis (3.21) even though advanced models exist (see in [60]), and are known to handle well transport mixing process.

$$-\bar{\rho}\widetilde{c''u''_j} = \frac{\mu_t}{\sigma_c}\frac{\partial(\tilde{c})}{\partial x_j} \tag{3.21}$$

where σ_c is the turbulent Schmidt number, here set to 0.7. Equation 3.20 contains the mean reaction rate term, $\widetilde{\dot{w}_c}$, which represents the classical mean reaction contribution common to turbulent single phase reacting flows. The extra contribution, $\widetilde{\dot{w}_{c,v}}$, is the droplet evaporation source due to droplet phase transition. These two reaction rate source contributions need modeling. Assuming that all droplets have evaporated before combustion the contribution of the evaporation source in (3.20) is neglected according to [35]. As simple approximation, we consider that the evaporation contribution is implicitly included through the changes in local mixture fraction field and turbulent flow parameters. Two different algebraic expressions of the mean reaction rate modeling are then used in this study. They are based on the flame surface density and on the assumption that the flamelet geometry has a fractal character according to [49]. Restricted first to premixed flame, we follow the Lindstedt and Váos formulation [50]:

$$\widetilde{\dot{w}_{c,v}} = \rho_u \frac{s_L^0}{V_K} C_L^{-2/3}\left(\frac{\varepsilon}{k}\right)\tilde{c}(1 - \tilde{c}) \tag{3.22}$$

where $V_K = (\varepsilon\nu)^{1/4}$ represents the Kolmogorov velocity and $C_L = 0.41$ at Re ≥ 50. The final expression for the mean reaction rate given in [50] is

$$\widetilde{\dot{w}_c} = C_R \rho_u \tilde{s}_L^0 \frac{\varepsilon}{k} \frac{1}{V_K} \tilde{c}(1 - \tilde{c}) \tag{3.23}$$

where the parameter C_R was turned to 2.6 to reproduce the measurements of a counter-flow premixed flame. Muppala and Dinkelacker [53] tracked the evolution of the pre-constant C_R that was set to 1.95 by Gulati and Driscoll [61] and to 1.35 by Videto and Santavicca [62]. It was turned to 1.25 by Abdel-Gayed et al. [63].

As the model has been formulated for ambient pressure, Muppala and Dinkelacker [53] considered an explicit pressure dependency and determined the parameter C_R as:

$$C_R = \frac{4.0}{Le} \left(\frac{p}{p_0} \right)^{0.5/Le^2} \tag{3.24}$$

where p denotes the operating pressure and p_0 the ambient pressure of 1 bar. Accounting explicitly for the fractal dimension, another form of Eq. 3.22 emerges as

$$\widetilde{\dot{w}_c} = \rho_u s_L^0 \left(C_L \frac{k^{3/2}}{\varepsilon} \right)^{D-3} \left(\frac{\nu^{3/4}}{\varepsilon^{1/4}} \right)^{2-D} \tilde{c}(1 - \tilde{c}) \tag{3.25}$$

The constant C_L is taken to 0.41 in all simulations and the fractal dimension D that should be within the theoretical range $7/3 \leq D \leq 8/3$ is set to 7.7/3. The parameter s_L^0 in Eqs. 3.22 and 3.25 denotes the laminar burning velocity.

To especially account for partially premixed features, we base on Eqs. 3.20–3.25 by now considering variable equivalence ratio with mixture fraction. This means, we consider that the laminar burning velocity, s_L^0, together with the expansion ratio, shall depend on the equivalence ratio, that in turn, can be directly related to the mixture fraction that takes zero value at pure oxydizer and unity at pure fuel. The first consideration of this effect in the BML theory was reported by Maltsev et al. [51, 52]. According to [52], we then extend the BML theory through a coupling with the mixing transport and the finite chemistry. In order to take into account the turbulence-chemistry interaction a presumed PDF approach is applied. The PDF is assumed to have a form of a Beta-function even though this is questionable in evaporating spray environments [15]. Thus, the laminar burning velocity and the expansion rate are function of the mean mixture fraction and its variance that are described by their respective transport equations (see Table 3.1).

$$s_L^0 = f\left(\tilde{z}, \widetilde{z''^2}\right); \quad \tau = f\left(\tilde{z}, \widetilde{z''^2}\right) \tag{3.26}$$

By doing so and in order to extract the physical products Y_α, results of conditioned combustion products must be multiplied by the probability p^b of being behind the flame front

$$Y_\alpha\left(z, p^b\right) = Y_\alpha^b(z) p^b = Y_\alpha^b(z) c \tag{3.27}$$

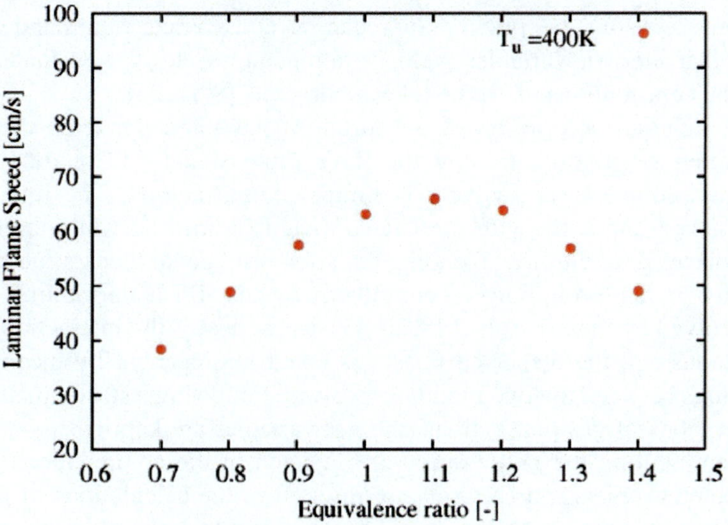

Fig. 3.1 Laminar flame speed of N-dodecane/air mixtures with unburned mixture temperature of 400 K [59]

with $p^b = c$ in the context of the BML model. Applying the bimodal PDF of c as assumed it can be shown that [52]

$$\tilde{Y}_\alpha\left(\widetilde{c},\widetilde{z},\widetilde{z''^2}\right) = \widetilde{c}\widetilde{Y_\alpha^b}\left(\widetilde{z},\widetilde{z''^2}\right) \tag{3.28}$$

The functional dependencies (3.26) and (3.28) that are commonly referred to as flamelet library are tabulated in a preprocessing step following [47] to speed up the numerical calculations and used as look-up table in the simulations. Thereby a reaction mechanism describing the kerosene chemistry has been required. Having in mind that the major components for kerosene are alkanes, aromatics and alkenes, kerosene has been substituted by N-dodecane following [64–66]. The detailed chemical reaction mechanism for N-dodecane involved 57 species and 281 reactions. To generate the flamelet tables, Lewis number was set to unity and the strain rate equaled 100/s. In particular, the laminar burning velocity for N-dodecane was obtained fitting the experimental data measured at the mixture preheat temperature of 400 K by Kumar and Sung [59]. The flammability limits are about 0.7 and 1.4, respectively, for the lean and rich compositions as shown in Fig. 3.1.

2.5 Numerical Procedure and Coupling

As the computational method is based on an Euler-Lagrangian coupling approach, two solvers have been involved and coupled. In the 3-D Lagrangian code (LAG3D) for particle tracking, the Lagrangian equations for droplets are discretized using first order scheme and solved explicitly. Source terms and average values and variances of droplet characteristic variables (velocity, temperature, etc.) are evaluated in each cell with the contributions of all the relevant droplets [8, 16, 23].

For the Eulerian description of the turbulent gas phase far from the droplet, the simulation is performed using the three dimensional CFD-code FASTEST in which the equations are solved by finite volume method. The time integration is achieved implicitly with the Crank-Nicholson method while the diffusion terms are discretized with flux blending on a non orthogonal block structured grid. The velocity-pressure coupling is accomplished by a SIMPLE algorithm. The whole system is solved by the SIP-solver [8, 23, 43]. Numerically, the interaction between the continuous and the dispersed phases is taken into account by means of several couplings between the two modules involved. Following a steady method, after several iterations of gas phase alone, the gas variables are kept frozen and all the droplets representing the entire spray are injected in the computational domain. The computed droplets source terms are inserted in the calculations of gas phase and kept frozen till the next coupling takes place in which the old particle sources are replaced by newly calculated sources. High levels of under-relaxation technique were used in order to obtain successful convergence [8, 23, 46]. Due to the presence of droplets source terms, the conventional residuals are characterized by a jump of residuals after each coupling. To avoid it, an additional under-relaxation technique should also be employed for droplet source terms [46].

$$S^{i+1}_{\phi\, p} = S^{i}_{\phi\, p} \cdot (1-\gamma) + S^{i+1}_{\phi\, p(cal.)} \cdot \gamma$$

where $S^{i+1}_{\phi\, p}$ and $S^{i}_{\phi\, p}$ are the particle/droplet source terms appearing at i + 1th and ith couplings, respectively. The under-relaxation factor γ takes values in the interval [0,1]. In the frame of this work, convergence of the Eulerian-Lagrangian coupling procedure is reached when the fluid's properties do not change their value from one coupling to the next in the presence of droplets.

3 Configurations

3.1 Geometry Description

The single annular combustor under study represents the experimental set up used by Baessler et al. [2] as reproduced in Fig. 3.2. The burner consists of two parts: the pre-vaporization zone and the combustion tube. The kerosene fuel is fed (1) to an ultrasonic nozzle (8), which uses an additional mass flow of the carrier air (2) of 20 ln/min (normalized l/min) for the improvement of the radial spray dispersion during the injection. The main air (3) is heated by a set of sinter metal plates (4) with an integrated heater coil (5). The main air volume flow rate was 300 ln/min and its

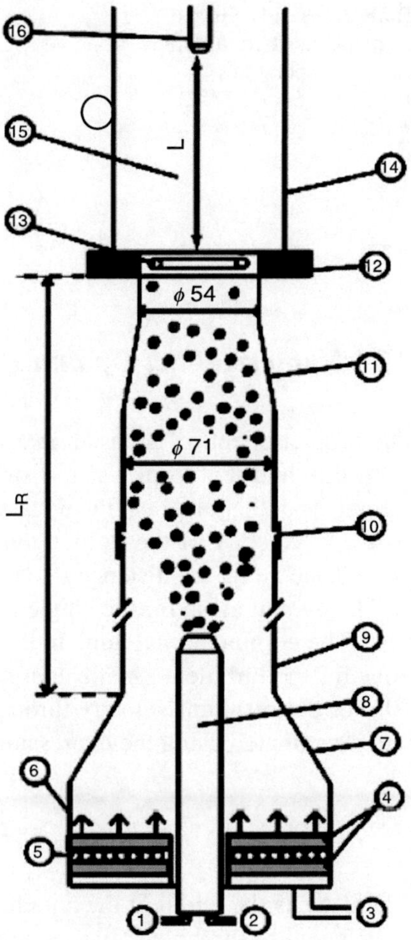

Fig. 3.2 Sketch of the experimental setup (partially premixed pre-vaporized combustor) [2]

temperature equals 90 °C. It enters the vessel (6) and is accelerated in a nozzle (7) before it enters the pre-vaporization zone (9). The droplets are being injected with a Sauter mean diameter of 50 μm and evaporate in the glass tube of 71 mm diameter. The length of the pre-vaporization zone L_R can be adjusted by adding tube elements which are interconnected using aluminum rings (10). Downstream the mixing zone, a second nozzle (11) of 54 mm diameter is used to increase the flow velocity and prevent flame flash back. The mixture is ignited by a hot wire ring (13) mounted in a water cooled ring (12). In the wake of this ignition source a stable flame develops and spreads over the cross section further downstream in the combustion zone (14). Exhaust gas measurements were taken at the tube axis at a distance H = L (15) (16) above the hot wire ring, while the PDA measurements were carried out across the tube section at length L_R. More details of the experimental set up and measurement techniques used can be found in [2].

Table 3.2 Grid cells and droplet number for different L_R and temperatures $T = 20\,°C, 50\,°C, 77\,°C, 90\,°C, 105\,°C$, respectively

Length [m]	Number CV [1,000]	Number particles [1,000]
0.5	450	160
0.6	475	160
0.7	500	160
0.8	522	160
0.9	560	160
1	650	160

3.2 *Numerical Set Up and Boundary Conditions*

The droplet injection is based on a stochastic approach by considering the droplet mass flux and the droplet size distributions obtained from the experimental measurements at the inlet near the nozzle exit. The kerosene droplets are initialized with 90% slip velocity of the fluid element. The mixing between the air and the vapor takes place along the distance L_R (Fig. 3.2).

The overall mesh for the single annular combustor is given in Table 3.2 for each case. The computational domain is exemplarily shown in Fig. 3.3. The inlet conditions for the turbulent kinetic energy are calculated using a turbulence intensity of 10% of the resultant velocity through the inlet. The distribution of the dissipation rate is estimated using the expression

$$\varepsilon = C_{\mu}^{3/4} \frac{k^{3/2}}{0.41 \cdot \Delta r} \tag{3.29}$$

where Δr is the width of the annulus. Here the turbulent length scale was assumed to be equal to the hole's diameter or inlet's opening. The mixture fraction boundary conditions are set to zero at all inlets ($\tilde{z} = 0$), since the injected air does not contain any fuel and the variation of mixture fraction is originated only by the produced vapor.

Since there was no temperature variation at the inflow, the progress variable was set to zero at all inlet boundaries except at the position of the hot wire ring for the ignition or stabilization and insurance of continuous combustion. The reference global equivalence ratio equals 0.7.

4 Results and Discussion

As operating condition for the reference simulation, the carrier phase temperature is set to 90 °C, and the length of the pre-vaporization zone equals 0.8 m. The number of the numerical droplets were 160,000 within one coupling. The summary of all computational cases classified with respect to physical parameters (tube length, air

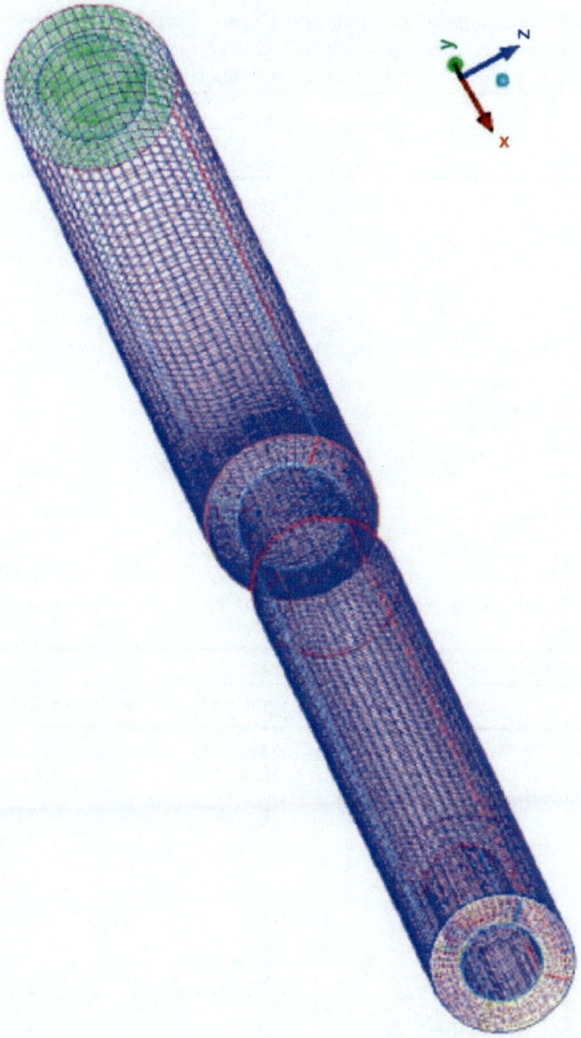

Fig. 3.3 Numerical grid

temperature, SMD), processes involved (ignition, non-equilibrium or equilibrium evaporation modeling) and model parameters varied is given in Tables 3.3 and 3.4. When plotted, the properties of the dispersed phase featured a smooth profile, i.e. the statistics were completely reliable. Increasing the droplet number did not change the results but enhanced the computing time.

Let us first focus on the spray process within the pre-vaporization zone (cases A in Table 3.3 and 3.4). Here the kerosene droplets are subjected to heated environment that leads to droplet evaporation. As the evaporation length is variable, it was changed from 0.5 to 1 m (see Tables 3.2). For each evaporation length tube, the spray and the heated air were mixed in the glass tube. The temperature of the

Table 3.3 Summary of computational cases classified with respect to physical parameters, processes involved and model parameters varied. (Only quantities that differed from the reference case Ref0 are indicated)

	Parameter (a)				
Case	D	ϕ	S_L	$\bar{S}_{\widetilde{z''^2},p}$	C Eq. 3.31
Ref0	7.7/3	0.7	$[0.7, 1.45.10^4]$	Eq. 3.34	2
A					
B					
C					
D					
E			$[0.6, 1.45.10^4]$		
F					
G	7/3, 8/3				
H				-Eq. 3.33 - Zero	
I					1

Table 3.4 Summary of computational cases classified with respect to physical parameters, processes involved and model parameters varied. (Only quantities that differed from the reference case Ref0 are indicated)

	Parameter (b)				
Case	L_R [m]	Evaporation model	Ignition location	SMD [μm]	T [°C]
Ref0	0.8	neq-model	yes (H < 0.2 m)	50	90
A	Table 3.2		No		Table 3.2
B	Table 3.2		Yes		
C		Eq-model			
D				20	
E					
F			- H > 0.2 m - Within L_R		
G					
H					
I					

mixing air and the length of the pre-heater tube were used for the control of the degree of vaporization. Figures 3.4–3.11 compare numerical and experimental results for different evaporation tube lengths at different mixing air temperatures. The measurement took place at the end of the tube length L_R (see Fig. 3.2).

4.1 Influence of Mixing Air Temperature and Pre-vaporization Tube Length on Vaporization Degree (Case A)

The first task was to produce the flow field, especially the droplet velocity at given mixing air temperature as function of tube lengths of the pre-vaporized zone.

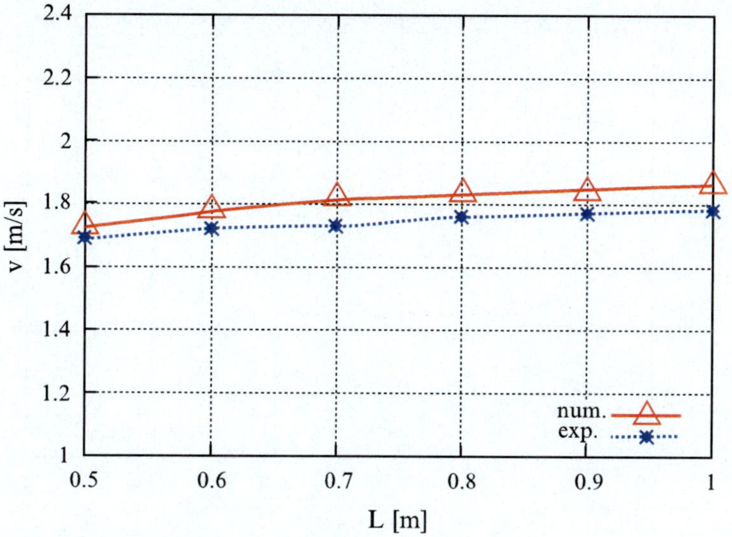

Fig. 3.4 Droplet axial velocity component at T = 50 °C

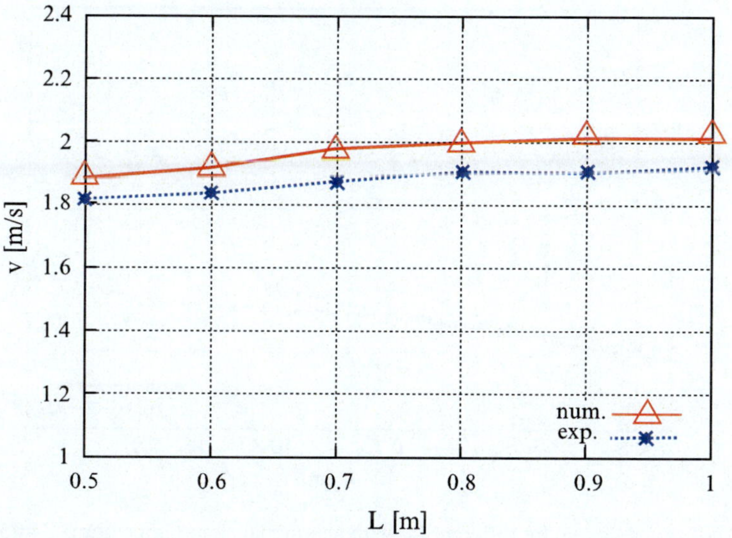

Fig. 3.5 Droplet axial velocity component at T = 77 °C

These results, that are also compared to experimental data, were achieved by using two different evaporation models, the equilibrium model (eq-model) and the non-equilibrium model (neq-model). The flow field results reported in Fig. 3.4 were obtained by using the neq-model for evaporation as it is shown below to describe well the evaporation process in the configuration under study.

A good prediction of the droplet axial velocity component once compared with experimental data is evident. Each data point represents a mean value on the

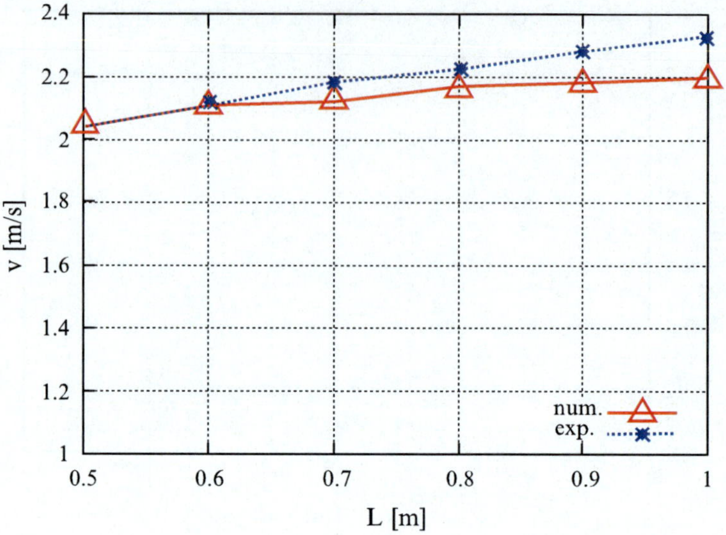

Fig. 3.6 Droplet axial velocity component at T = 105 °C

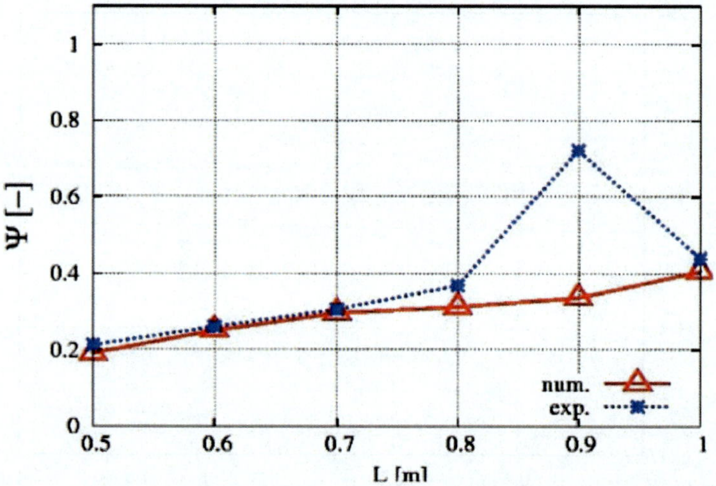

Fig. 3.7 Vaporization degree for different L_R with neq-model at air temperature T = 50 °C

measurement point of the axial axis at the end of the tube length. The overall slight deviation may be due to inflow conditions where breakup process has not been considered. With this confidence, further properties could be retrieved. Figures 3.7–3.11 show the evolution of the evaporation degree of kerosene fuel along the evaporation tube at different mixing air temperatures. Thereby, results obtained by using both evaporation models are compared with experimental data.

The results in Figs. 3.7–3.10 show clearly that the evaporation degree is getting higher both increasing the pre-vaporization zone length and the inlet air temperature.

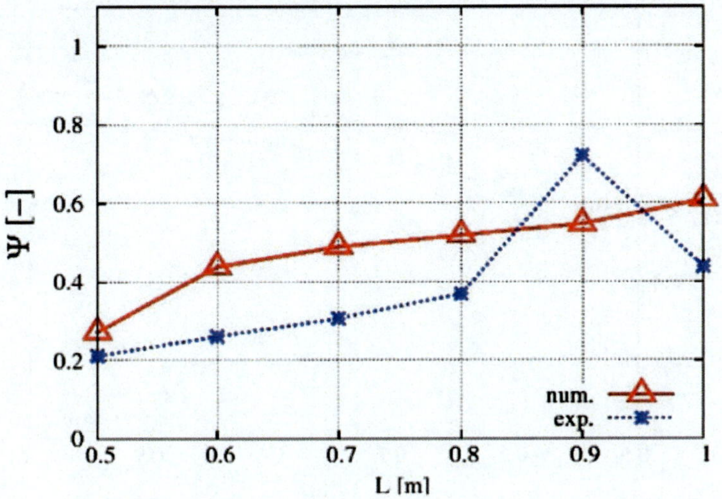

Fig. 3.8 Vaporization degree for different L_R with eq-model at air temperature T = 50 °C

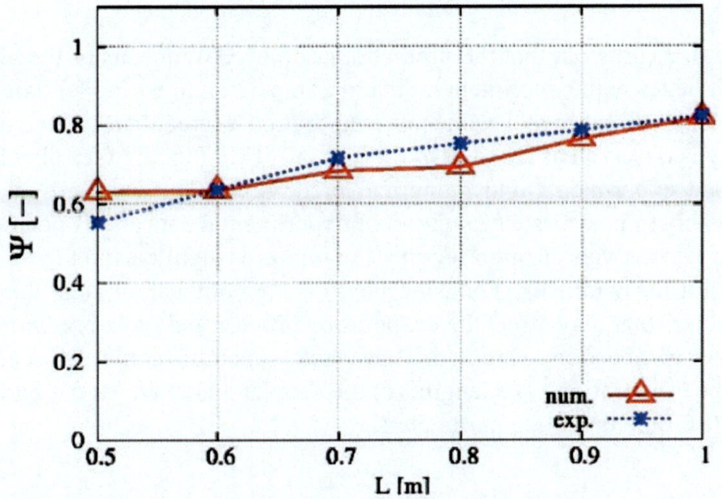

Fig. 3.9 Prevaporization degree for different L_R with neq-model at air temperature T = 77 °C

This is summarized in Fig. 3.11 where only results obtained by using the neq-model are reported. According to experiments, the evaporation degree was defined as:

$$\Psi = \left. \frac{\dot{m}_{ref(T=20°C)} - \dot{m}_T}{\dot{m}_{ref(T=20°C)}} \right|_L \tag{3.30}$$

at a given axial section located at a distance L from the nozzle exit. In this relation $\dot{m}_T$ expresses the mass flux at temperature T and $\dot{m}_{ref(T=20\,°C)}$ is at $T = 20\,°C$ when no evaporation is present.

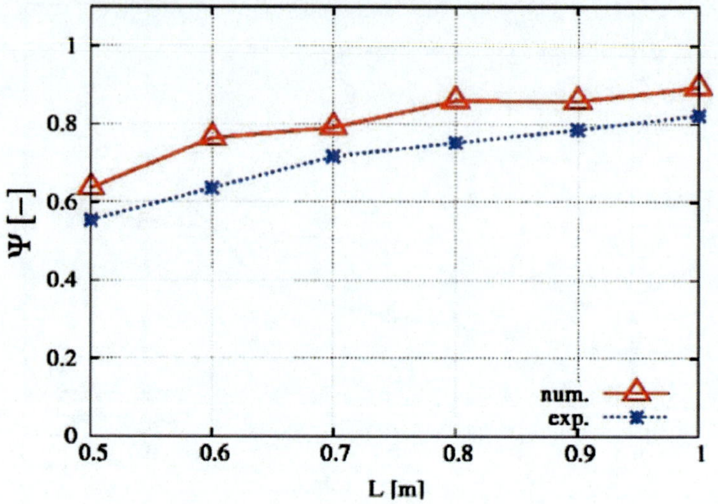

Fig. 3.10 Vaporization degree for different L_R with eq-model at air temperature $T = 77\,°C$

In general, it turns out that the predictions achieved by means of the neq-model agree much better with experimental data in comparison to eq-model based predictions. The latter reveal higher values of evaporation degree than expected. A large discrepancy is observed at $L = 0.9\,m$ for $T = 50\,°C$ in Fig. 3.7 (see also Fig. 3.11, left) between experiments and numerical predictions. This isolated case can be linked probably to inaccurate evaluation of experimental data at this position.

Downstream the vaporization section, the mixture is ignited and the flame is stabilized with a hot wire ring. Focusing now on the combustion zone the effect of different factors that may affect the evaporation process and/or its prediction, and in turn the spray combustion, is studied. These factors include (a) the evaporation modeling, (b) the evaporation zone length, (c) the droplet diameter, (d) the flammability limits and (e) some modeling parameters.

4.2 Influence of Evaporation Models (Case C)

A qualitative view of the temperature distribution along the axis of the combustion tube as predicted by using the neq-model is shown in Fig. 3.13. Thereby $Y = 0.8{+}H$. In Fig. 3.12 "0.0 m" represents the position of the ignition source. Note that the mixing is mainly influenced by the degree of evaporation of the kerosene droplets, which, in turn, is affected by the carrier phase temperature and the length of the pre-vaporization zone. If the eq-model is used, higher values of evaporation degree in the pre-vaporized zone is predicted, as depicted in Figs. 3.7–3.10. This leads to an early formation of an ignitable mixture in the combustion zone and

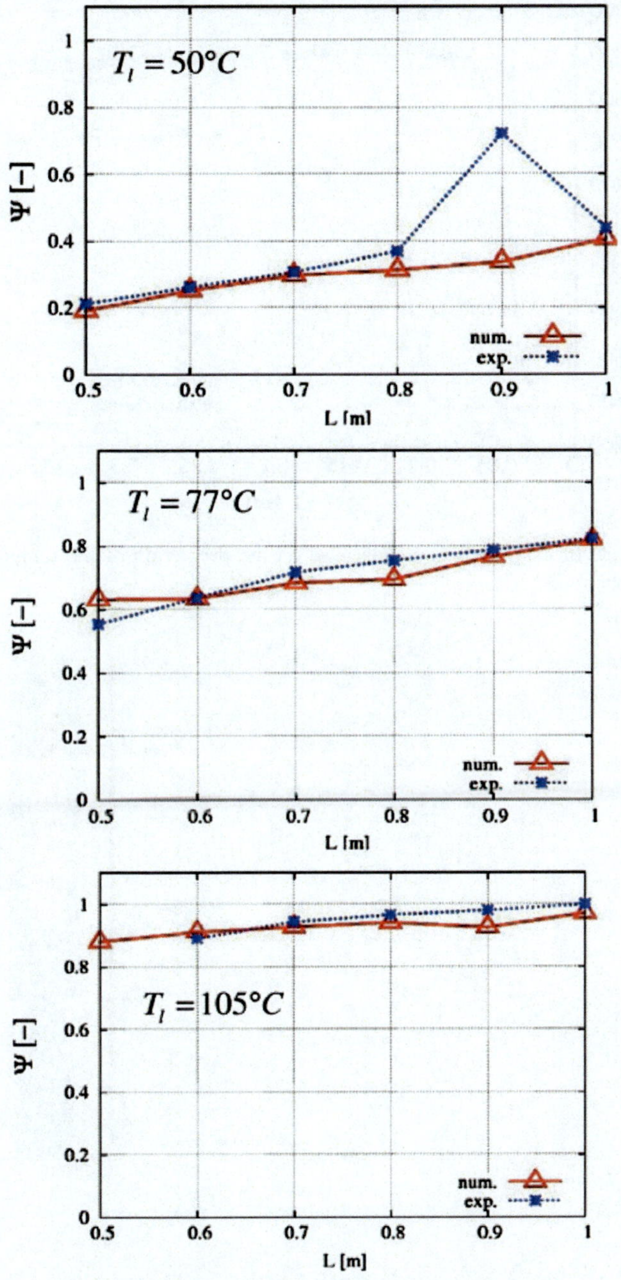

Fig. 3.11 Vaporization degree for different L_R at fixed inlet temperature

results in a predicted combustion that takes place 0.075 m earlier than that predicted by the neq-model. It appears that the experimental data for the temperature is better approximated by the use of neq-model. In particular the flame lift-off agrees very well with the experimental measurements, whereas the temperature maximum

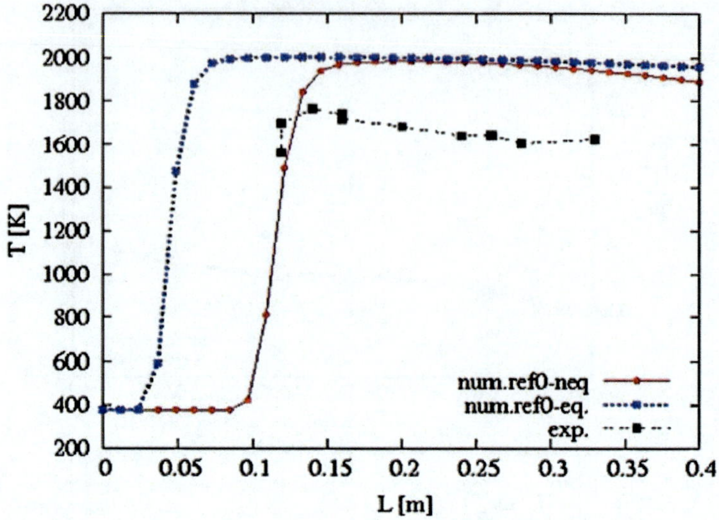

Fig. 3.12 Influence of evaporation models used on the prediction of the temperature along the combustion chamber axis

Fig. 3.13 Temperature distribution as calculated using neq-model. Y = 0.8 + H

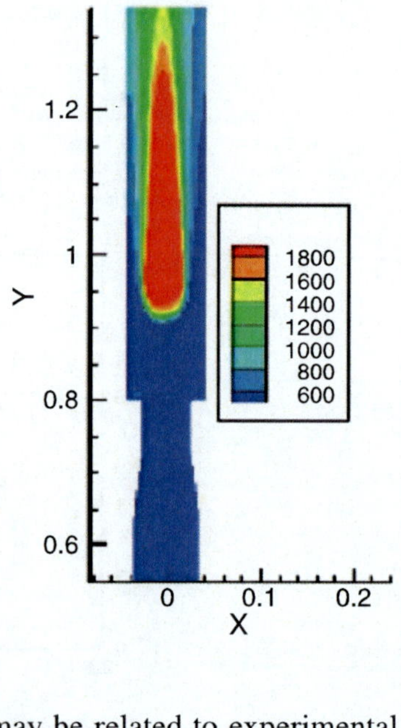

values show a ΔT of ca. 250 K. This discrepancy may be related to experimental data, since the temperature analysis showed a very high fluctuation frequency (of ca. 450 K) which is a typical limitation for thermocouple measurements.

The lift-off occurrence is confirmed in Figs. 3.14 and 3.15 that show qualitatively the field of the mixture fraction obtained with both evaporation models. The stoichiometric contour predicted by the neq-model corresponds to a value of 0.063. A lean mixture is observed between the kerosene vapor and the oxidizer

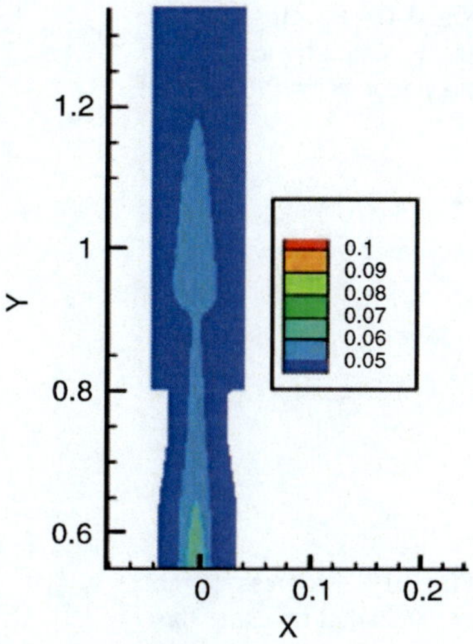

Fig. 3.14 Mixture fraction as calculated by using the neq-model. Y = 0.8 + H

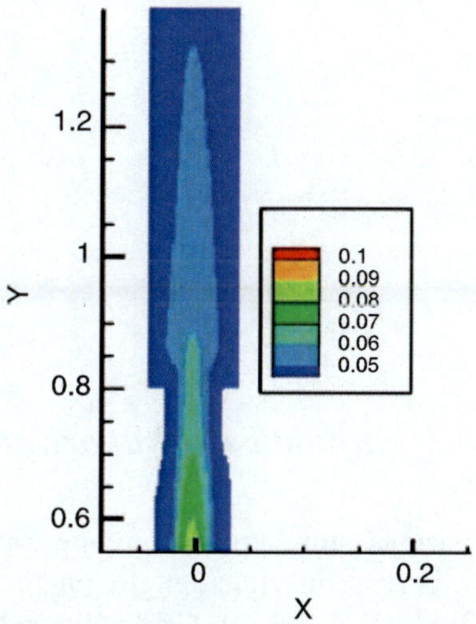

Fig. 3.15 Mixture fraction as calculated by using the for eq-model. Y = 0.8 + H

in the second part of the combustion tube. The mixing is concentrated around the centerline of the configuration; close to the wall the mixture fraction equals zero. This is due to the dispersion and entrainment of the dispersed phase. The progress variable, that distinguishes between the fields of burned and unburned gases, is presented in Figs. 3.16 and 3.17. The form and position of the flame front are clearly observable. The maximum values of the progress variable are located along the axis; the flame lift-off that could be easily quantified by the flexion point of the temperature profile in Fig. 3.12 equals ca. 0.12 m. The progress variable has zero values in the vicinity of the wall because of the poor mixing in the region, i.e. the vapor concentration is very low and the mixture lies out of the lower flammability domain.

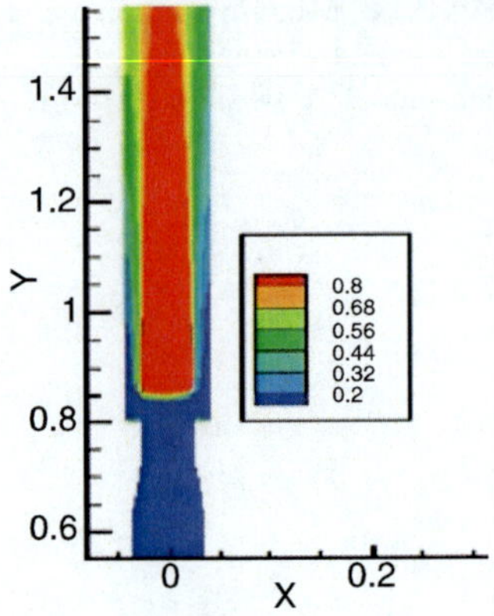

Fig. 3.16 Progress variable as calculated by using the eq-model. Y = 0.8 + H

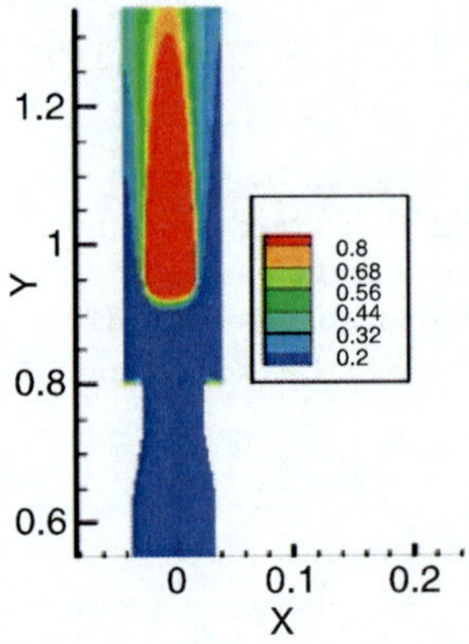

Fig. 3.17 Progress variable as calculated by using the neq-model. Y = 0.8 + H

4.3 Influence of Vaporization Zone Length (Case B)

Various simulations were performed to study the influence of varying the pre-vaporization zone length on the combustion process. They are summarized in Tables 3.3 and 3.4. Grid cells and tracked droplet numbers are plotted in Table 3.2. Figure 3.18 shows the temperature profile for different pre-vaporization lengths. The flame lift-off clearly diminished with decreasing values of L_R. A reduction of $\Delta L = 25\%$ reduced flame lift-off by $\Delta h = 0.05$ m. Due to the reduction of the pre-vaporization tube length, droplets have less time to stay there, thus the evaporation degree is in turn reduced. In contrast, the mass of kerosene droplets, which arrive at the combustion tube and evaporate or burn there, increases. This phenomenon influences the mixing process very strongly and a new field of mixture is obtained. The first part of the configuration exhibits a less rich mixture whereas close to the ignition source, the concentration of vapor increases and the mixture pushes the flammability limit. Thus the combustion process starts earlier than the reference one. Further, using a shorter pre-vaporization zone increases the maximum temperature by $\Delta T = 100$ K. With a decrease in the degree of vaporization, emissions are expected to become higher due to the increasing number and size of burning droplets that have to act as hot spots [2].

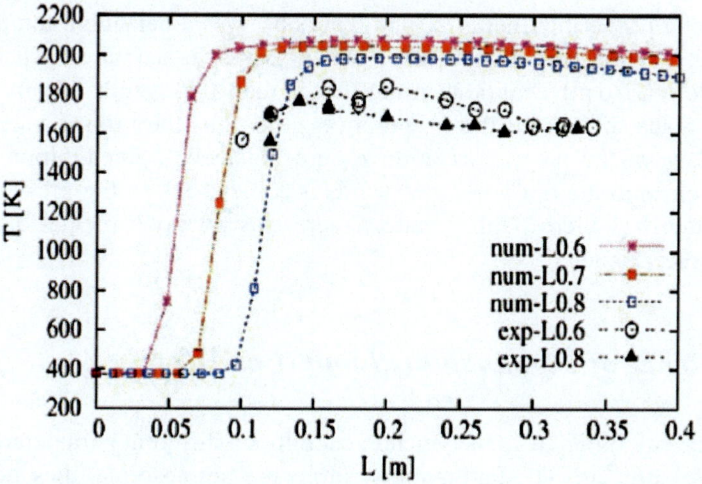

Fig. 3.18 Comparison of temperature profile for various pre-vaporization tube lengths

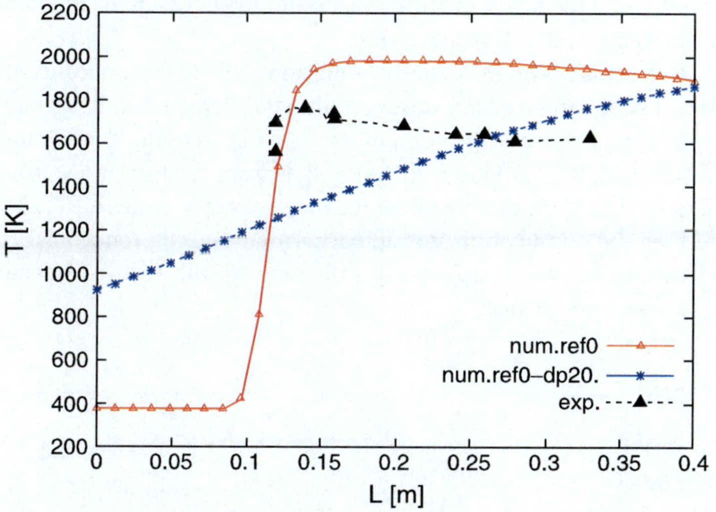

Fig. 3.19 Comparison of temperature profiles for droplet diameter 50 μm (ref0), 20 μm (dp20) and exp. (50 μm)

4.4 *Influence of Droplet Diameter (Case D)*

The boundary conditions provided by the measurements are limited for droplet spray with a Sauter mean diameter of 50 μm. The question is what would happen if droplets change the size distribution? The diameter size variations can be achieved by changing the geometry of the nozzle or even by changing the operating conditions. In gas turbine the Sauter mean diameter has often a value of 20 μm [17, 23]. Therefore simulations with a relatively smaller diameter than 50 μm are worth. Figure 3.19 presents the temperature profile for the case of spray with a Sauter mean diameter at the inlet equal 20 μm. One observes a very high temperature at the beginning of the combustion tube, which expresses a flame flash back. The small droplets (20 μm) evaporate much faster than these with 50 μm. They are not able to reach the end of the evaporation zone, thus the mixing takes place in the early stage of the pre-vaporization zone exclusively. After ignition the flame propagates close to the hot wire, afterwards it comes back to the pre-vaporization zone and stabilizes there. Unfortunately there was no experimental data for this class of droplet diameter.

4.5 Influence of Flammability Limit (Case E)

The flammability limits of fuels depend generally on different parameters e.g. temperature, pressure, etc. The flammability limits are not absolute, they depend also on the type and strength of the ignition source; therefore a slight variation on the flammability limits of dodecane, which feature different volatility of kerosene is worth to investigate. Following [59] the upper and lower limits of the kerosene surrogate were set to 0.7 and 1.4, respectively.

Figure 3.20 illustrates the influence of enlarging of the flammability limits from 0.7–1.4 to 0.6–1.4. It can be easily observed that the flame lift-off has been reduced with $\Delta h = 0.02$ m, whereas the temperature profile remained unchanged. Since the new laminar velocity has values different than zero for mixture starting from an equivalence ratio equal 0.6, the chemical reaction was able to arise for a leaner composition and then the combustion started earlier than the reference case. The upper flammability limit did not manifest any influence on the results, because we are dealing with a lean combustion.

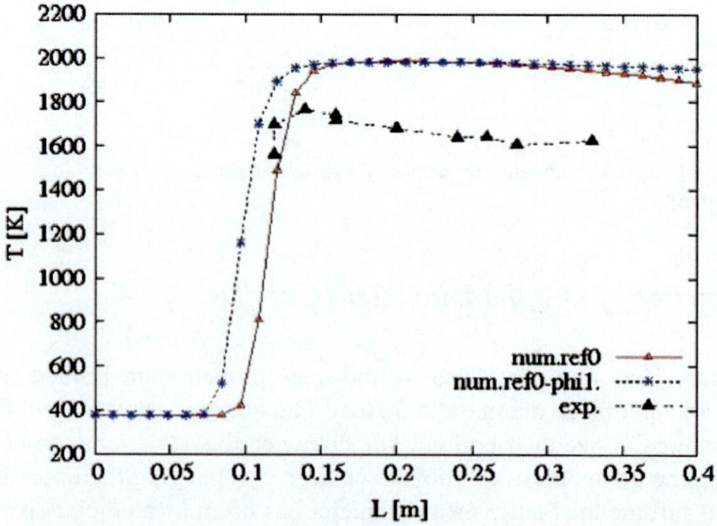

Fig. 3.20 Effects of the flammability limit enlarging (ref0-phi1) on the temperature distribution

This result attests the ability of the designed tool to well treat the process ongoing as it is in agreement with the findings in [59].

4.6 *Influence of Ignition Location (Case F)*

Ignition is a crucial phase in many practical applications. Instead of providing continuous ignition using a heat source (which was numerically performed by setting the boundary condition for the progress variable to 1 at the inlet close to the hot wire ring), the aspect using a spark ignition is investigated here. This subject is an important key for some design purposes especially relight of an aircraft gas turbine at high altitude, flame flashback, etc. As the spark position influences the ignition probability defined as successful flame establishment, three localizations for the initialization of the progress variable were chosen, as depicted in Fig. 3.21. For that purpose the boundary condition of the progress variable is set to zero in all inlets.

Two observations have been made. (1) A flame flashback occurs if the progress variable was initialized behind the hot wire, i.e. within the pre-vaporization zone. (2) The combustion process did not take place if the progress was initialized at a position higher than 0.2 m forward of the hot wire (H > 0.2 m). Figures 3.22 and 3.23 show the temperature profiles for the three cases of initialization. When flashback happens, the temperature rises up to 1,400 K at 0.0 m (Fig. 3.22). Figure 3.24 displays the progress variable field in a center plane cross-section. Due to the high

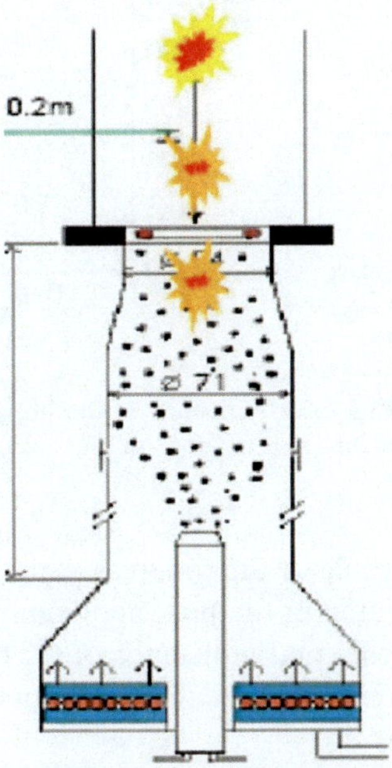

Fig. 3.21 Three positions for the combustion initialization

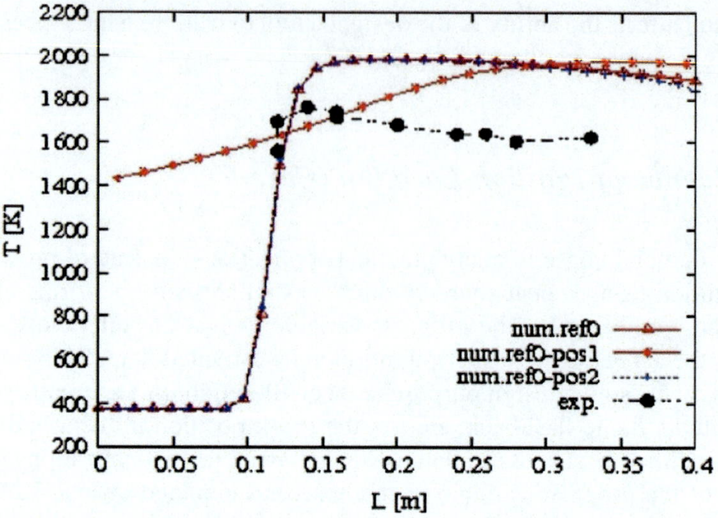

Fig. 3.22 Influence of the initialization location on the combustion process. ref0: continuous ignition, ref0-pos1: in the pre-vaporization zone, ref0-pos 2 = at H < 0.2 m

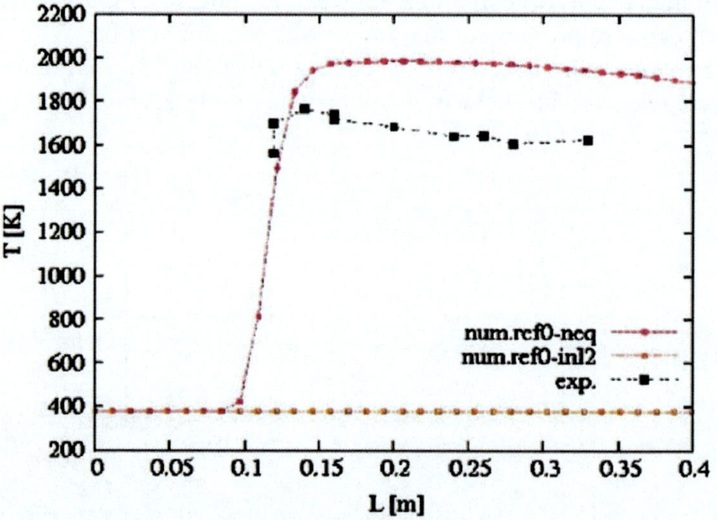

Fig. 3.23 Influence of the initialization location on the combustion process. ref0- neq: at H < 0.2 m (also ref-pos2 in Fig. 3.22), ref0-inl2: at H > 0.2 m

value of the progress variable, the flame flashback, and thus the temperature, both jump in the pre-vaporization zone. When the initialization occurs between the hot wire and a distance of 0.2 m forwards (H < 0.2 m), there was no difference with respect to the reference simulations.

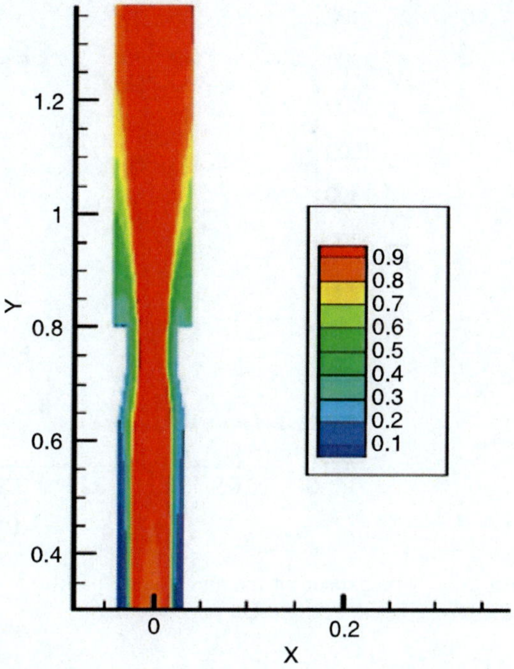

Fig. 3.24 Progress variable as calculated by using the neq-model

4.7 Influence of Model Parameters

Model parameters are also known to affect the predictions of numerical simulations. In the following this aspect is addressed by considering the impact of (1) the fractal parameters, and (2) the modeling approaches for the evaporation source term in the mixture fraction variance equation.

4.7.1 The Fractal Dimension D (Case G)

The fractal theory [48–52] was applied to model the reaction rate term in Eq. 3.22, which represents the most crucial part of the BML model. Thereby the fractal dimension D is adjusted for each flame within the theoretical range of $7/3 \le D \le 8/3$. If the surface is smooth, the fractal dimension, D, will approach 2. For rough surfaces, D will approach 3 while the surface appears to fill the volume of the cell under consideration.

The theoretical boundaries of D were numerically studied to verify how the model is sensitive to this parameter variation. Figure 3.25 displays the effects of the model parameter D on the temperature profile prediction. By using larger values for D it appears clearly that the flame lift-off shall be reduced, whereas for values of D smaller than the reference one, the flame lift-off increases. At the lower limit ($D = 7/3$), the flame becomes instable and could even extinct. By increasing the

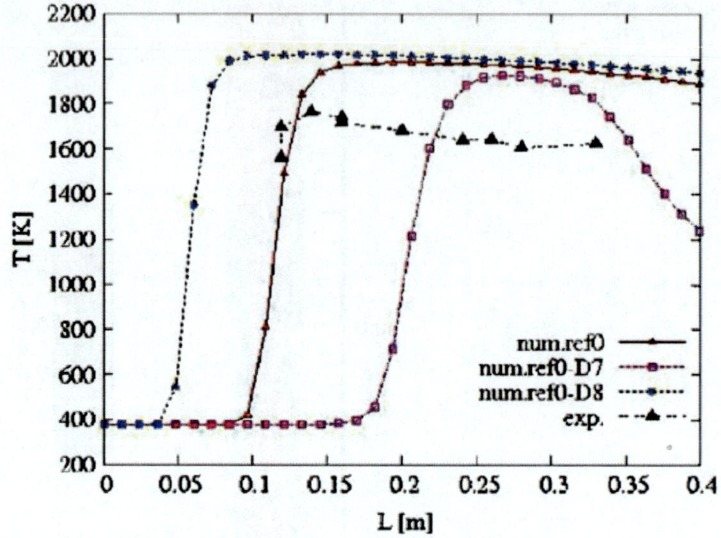

Fig. 3.25 Comparison of the temperature profiles at different values of D: ref0 for D $= 7.7/3$; ref0-D7 for D $= 7/3$; ref0-D8 for D $= 8/3$

fractal dimension, the flame surface will get rougher, i.e. the chemical reaction has more area within the same volume, making the reaction to appear more active and durable. By decreasing D, the flame decreases per volume unit and the chemical reaction becomes very limited so that the production of the reaction rate term reduces and the flame blows out. Both fractal dimension limits influence the results greatly.

4.7.2 The Modeling of the Evaporation Source Term in the Mixture Fraction Variance Equation (Case H)

The reliability of spray combustion models lies primarily on how the fuel-air mixture preparation is accurately described. It is then of interest to better understand the behavior of mixture fraction variance when spray evaporation is present. For this purpose, we focus on the source term in the transport equation of the Favre-averaged mixture fraction variance and investigate how this term may influence the prediction of the fluctuation level of fuel mass along with the mixture fraction and spray combustion properties prediction. This equation reads [15]:

$$\frac{\partial\left(\bar{\rho}\widetilde{z''^2}\right)}{\partial t}+\frac{\partial\left(\bar{\rho}\tilde{u}_j\widetilde{z''^2}\right)}{\partial x_j}=-\frac{\partial}{\partial x_j}\left(\bar{\rho}\frac{\upsilon_t}{S_{ct1}}\frac{\partial\widetilde{z''^2}}{\partial x_j}\right)+2\bar{\rho}\frac{\upsilon_t}{S_{ct2}}\left(\frac{\partial\tilde{z}}{\partial x_j}\right)^2$$
$$-c_Z\bar{\rho}\frac{\tilde{\varepsilon}}{\tilde{k}}\widetilde{z''^2}+\bar{S}_{\widetilde{z''^2},p} \tag{3.31}$$

The first term on the right hand side in (3.31) expresses the turbulent transport, the second represents the scalar dissipation rate while the third denotes the production/sink term. The model coefficients S_{ct1} and S_{ct2} are set equal to 0.7 according to [15]. The last term in (3.31) expresses the spray evaporation source term. According to [11] this term can be rewritten and formulated as

$$\bar{S}_{\widetilde{z''^2},p} = 2\widetilde{z'' \dot{m}_p} - 2\tilde{z}\widetilde{z'' \dot{m}_p} - \widetilde{z''^2 \dot{m}_p} = 2\widetilde{z'' \dot{m}_p}(1-\tilde{z}) + (-\widetilde{z''^2 \dot{m}_p}) \quad (3.32)$$

Different expressions exist in the literature for the modeling of this term. Subramanian et al. [67] tested three different closures in a combustion cell where the liquid fuel was injected through a high pressure, single hole nozzle, common-rail, Diesel type injector. In the present work two models are addressed. The first approach is the formulation by Hollmann and Gutheil [15] given as

$$\bar{S}_{\widetilde{z''^2},p} = \widetilde{z''^2} \tilde{\dot{m}}_p (1 - 2\tilde{z})/\tilde{z} \quad (3.33)$$

The second consists in the so-called single droplet model (SDM) by Reveillon and Vervisch [11] formulated as:

$$\bar{S}_{\widetilde{z''^2},p} = 2(1-\tilde{z}) \int_0^{z^s} (z^+ - \tilde{z})(\widetilde{\dot{m}_p | \tilde{z}}) P(z^+) dz^+ + \left(- \int_0^{z^s} (z^+ - \tilde{z})^2 (\widetilde{\dot{m}_p | \tilde{z}}) P(z^+) dz^+ \right) \quad (3.34)$$

where the conditional source term depends on the statistics of both the droplet size and the Spalding number and is approximated as a monotonic function of the mixture fraction, z:

$$(\widetilde{\dot{m}_p | \tilde{z}}) = \alpha_{B_M} z^n \quad (3.35)$$

Thereby the parameters α_{B_M} and n are determined as suggested in [11]. The probability $P(z^+)$ is assumed to follow a Beta distribution. More details about the model description can be found in [11]. The impacts of these models on the prediction of spray combustion properties, especially on the temperature profile are compared, as shown in Fig. 3.26.

Thereby simulations have been carried out using:

1. The model by Hollmann and Gutheil [15], labeled model 1 or ref0-varz1
2. The approximation by Reveillon and Vervisch [11], labeled model 2 or ref0-varz2,
3. No source term, quoted here as ref0.

The results in Fig. 3.26 show a slight difference of prediction only at the beginning of the tube. At $L = 0.15$ m the model 2 predicts a $\Delta T = 200°K$ whilst the model 1 calculates $\Delta T = 220°K$. Without considering the evaporation source term, the simulation achieves a $\Delta T = 240°K$. It turns out that the model by Reveillon

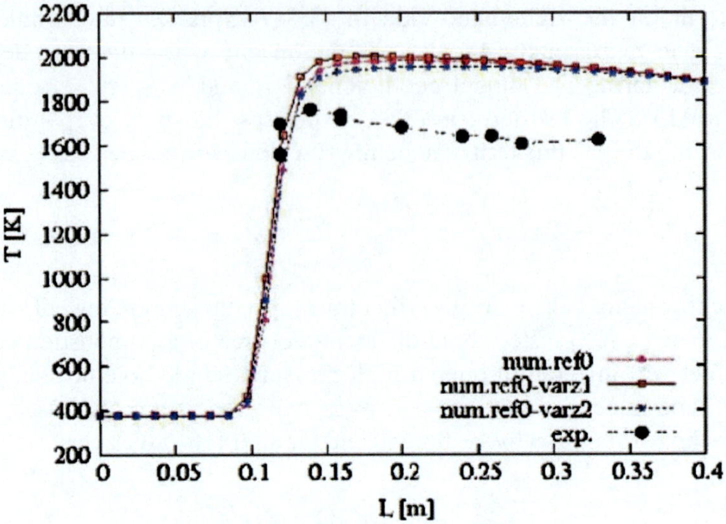

Fig. 3.26 Comparison of the temperature profiles for different modeling of variance source term: ref0: without source term; ref0-varz1: Gutheil; ref0-varz2: Vervisch

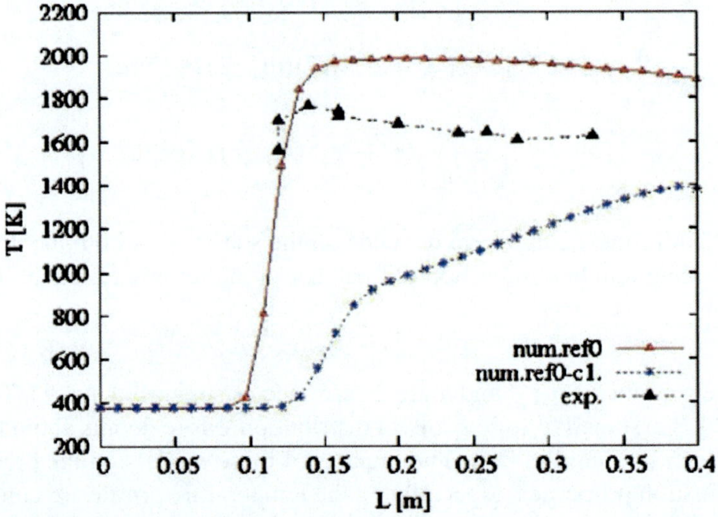

Fig. 3.27 Comparison of the temperature profiles for different values of the model parameter: c = 2: ref0; c =: ref0-c1

and Vervisch seems to be more accurate up to L = 0.3 m. The reason may be that the model 2 considers the joint statistics of the mixture fraction and the evaporation source. From the distance L = 0.3 m no difference can be observed, since almost all droplets are already evaporated. It appears clearly here that the evaporation source term in Eq. 3.31 cannot be neglected. This term influences essentially the fluctuation level in a limited zone close to the spray.

4.7.3 The Parameter C_Z in the Production/Sink Term (Case I)

Concentrated on the effect of the model parameter c_Z in the dissipation term of the variance equation (3.31), the results in Fig. 3.27 shows a relevant influence of this parameter on the prediction of the spray temperature. In the case $c_Z = 1$ the dissipation of the variance is small making the variance itself higher, that, in turn, leads to low temperature values. The case $c_Z = 2$ enables to capture better all spray combustion properties, especially the temperature profile. This value has been used as reference value in all the results reported throughout this paper. It also corresponds to the value used by Subramanian et al. [67].

5 Concluding Remarks

This study aimed at showing numerically the influence of droplet evaporation on the evolution of a partially premixed kerosene spray combustion in a model gas turbine combustor. It was possible to simulate turbulent spray combustion using an Eulerian-Lagrangian RANS based procedure under a full two-way coupling by consistently coupling different advanced sub-models. In particular, the combustion has been described by the so-called conditioned progress variable approach (CPVA) based on the Bray-Moss-Libby (BML) model which was extended to account for partially premixed combustion in LPP concept.

To assess the numerical approach simulation results and experimental data have been considered and compared. Based on the agreement achieved we confidently assess that the modeling approach performs well and a predictive capability is approaching once the evaporation is described by the non-equilibrium model. In particular,

- It could be shown from the comparisons between the equilibrium evaporation model and non-equilibrium evaporation model that non-equilibrium effects are important in case of droplet sizes typical for practical gas turbine spray configurations.
- The influence of both operating parameters, the mixing air temperature and the vaporization tube length, on the degree of vaporization and subsequently on spray flame characteristics has been correctly captured.
- The effects of various physical and modeling parameters on the prediction of the combustion were retrieved satisfactory:
  - The influence of droplet diameter, flammability limits and location of the combustion initialization has been pointed out
  - The effect of modeling parameters, like fractal dimensions in the expression describing the reaction chemical source term, the evaporation source term and the sink term in the modeled equation of the mixture fraction variance, has been found to be relevant.

Although consistent results with experimental data have been achieved in this work, using models of gaseous combustion to describe spray flames arise some modeling issues. The influence of droplet evaporation source in flamelet based modelling, the role of group combustion and of a flame index characterization have to be well addressed.

In this study, the effects of critical pressure conditions and the problems associated with atomization, primary and secondary break-up as well as droplet-droplet interaction have not been considered. These should be addressed in future work.

Acknowledgment The authors acknowledge the financial support from the German Research Council (DFG) through the SFB568 and GRK1144.

References

1. Ramotowski, M.J., Roby, R.J., Leo, P.E., Eskin, D., Klassen, M.S.: Fuel flexibility for dry low emission gas turbines – cleanly burning biofuels, coal liquids and petroleum fuels. PowerGen Conference and Exposition 2007, New Orleans, LA, USA, 10–13 December (2007)
2. Baessler, S., Moesl, K., Sattelmayer, T.: NOx emissions of a premixed partially vaporized kerosene spray flame. J. Eng. Gas Turb. Power **129**, 695–702 (2007)
3. Schäfer, O., Koch, R., Wittig, S.: Flashback in lean prevaporized premixed combustion: non swirling turbulent pipe flow study. J. Eng. Gas Turb. Power **125**, 670–676 (2003)
4. Rokke, N.A., Wilson, A.J.: Experimental and theoretical studies of a novel venturi lean premixed prevaporized (LPP) combustor. J. Eng. Gas Turb. Power **123**, 567–573 (2001)
5. Nomura, H., Hayasaki, M., Ujiie, Y.: Effects of fine fuel droplets on a laminar flame stabilized in a partially prevaporized spray stream. Proc. Combus. Inst. **31**(2), 2265–2272 (2007)
6. Balachandar, S., Eaton, J.K.: Turbulent dispersed multiphase flows. Annu. Rev. Fluid Mech. **42**, 111–133 (2010)
7. Ahmadi, G., Cao, J., Schneider, L., Sadiki, A.: A thermodynamically formulation for chemically active multiphase turbulent flows. Int. J. Eng. Sci. **44**, 699–720 (2006)
8. Chrigui, M., Ahmadi, G., Sadiki, A.: Study on interaction in spray between evaporating droplets and turbulence using second-order turbulence RANS models and a Langrangian approach. Prog. Comput. Fluid Dyn. **4**(3–5), 162–174 (2004)
9. Sazhin, S.S.: Advanced models for fuel droplet heating and evaporation. Prog. Energy Combust. Sci. **32**, 162–214 (2006)
10. Luo, K., Desjardins, O., Pitsch, H.: DNS of droplet evaporation and combustion in a swirling combustor. Center for Turbulence Research, Annual Research Briefs, pp. 253–265 (2008)
11. Reveillon, J., Vervisch, L.: Spray vaporization in non-premixed turbulent combustion modeling: a single droplet model. Combust. Flame **121**, 75–90 (2000)
12. Bellan, J., Selle, L.C.: Large eddy simulation composition equations for sigle-phase and two-phase fully multicomponent flows. Proc. Combust. Inst. **32**(2), 2239–2246 (2009)
13. Sankaran, V., Menon, S.: LES of spray combustion in swirling flows. J. Turbul. **3**(1), 11–11(1), (2002)
14. Apte, S.V., Mahesh, K., Moin, P.: Large-eddy simulation of evaporating spray in a coaxial combustor. Proc. Combust. Inst. **32**(2), 2247–2256 (2009)
15. Hollmann, C., Gutheil, E.: Modeling of turbulent spray diffusion flames including detailed chemistry. Proc. Combust. Inst. **26**(1), 1731–1738 (1996)
16. Chrigui, M., Moesl, K., Ahmadi, W., Sadiki, A., Janicka, J.: Partially premixed prevaporized kerosene spray combustion in turbulent flow. Exp. Therm. Fluid Sci. **34**(3), 308–315 (2010)
17. Smiljanovski, V., Brehm, N.: CFD liquid spray combustion analysis of a single annular gas turbine combustor, ASME paper 99-GT-300 (1999)

18. Berlemont, A., Grancher, M.S., Gouesbet, G.: Heat and mass transfer coupling between vaporizing droplets and turbulence using a Lagrangian approach. J. Heat Mass Transf. **38**, 3023–3034 (1995)
19. Miller, R.S., Harstad, K., Bellan, J.: Evaluation of equilibrium and non-equilibrium evaporation models for many droplet gas liquid flow simulation. Int. J. Multiphas. Flow **24**, 1025 (1998)
20. Bellan, J., Sommerfeld, M.: Theoretical examination of assumptions commonly used for the gas phase surrounding a burning droplet. Combust. Flame **33**, 107–122 (1978)
21. Jackson, R., Davidson, B.J.: An equation set for non-equilibrium two-phase flow, and an analysis of some aspects of choking, acoustic propagation, and losses in low pressure wet steam. Int. J. Multiphas. Flow **9**(5), 491–510 (1983)
22. Sornek, R.J., Dobashi, R., Hirano, T.: Effect of turbulence on vaporization, mixing, and combustion of liquid-fuel sprays. Combust. Flame **120**, 479–491 (2000)
23. Sadiki, A., Chrigui, M., Janicka, J., Maneshkarimi, M.R.: Modeling and simulation of effects of turbulence on vaporization, mixing and combustion of liquid-fuel sprays. Flow Turbul. Combust. **75**, 105–130 (2005)
24. Abramzon, B., Sirignano, W.A.: Droplet vaporization model for spray combustion calculations. Int. J. Heat Mass Transfer **32**(9), 1605–1618 (1989)
25. Aggarawal, S.K., Tong, A.Y., Sirignano, W.A.: A comparison of vaporization models in spray calculations. AIAA J. **22**(10), 1448–1457 (1984)
26. Godsave, G.A.E.: Studies of the combustion of drops in a fuel spray: the burning of single drops of fuel. In: Proceedings of the Fourth Symposium (International) on Combusion, Combustion Institute, Baltimore, pp. 818–830 (1953)
27. Spalding, D.B.: The combustion of liquid fuels. In: Proceedings of the Fourth Symposium (International) on Combustion, Combustion Institute, Baltimore, pp. 847–864 (1953)
28. Hubbard, G.L., Denny, V.E., Mils, A.F.: Droplet evaporation: effects of transients and variable properties. Int. J. Heat Mass Transfer **18**, 1003–1008 (1975)
29. Oefelein, J.C., Aggarwal, S.K.: Toward a unified high pressure drop model for spray simulation, center for turbulence research. In: Proceedings of the Summer Program 2000, Stanford University, pp. 193–205 (2000)
30. Becker, J., Hassa, C.: Breakup and atomization of a kerosene jet in cross flow of air at elevated pressure. In: ILASS-Europe 99, Toulouse (1999)
31. Park, J.H., Yoon, Y., Hwang, S.S.: Improved tab-model for prediction of spray droplet deformation and breakup. Atomization Spray **12**, 387–402 (2002)
32. Chui, H.H., Liu, T.M.: Group combustion of liquid droplets. Combust. Sci. Technol. **17**, 127–142 (1977)
33. Gutheil, E.: Modeling and simulation of droplet and spray combustion. Handbook of Combustion. Wiley-VCH Verlag GmbH & Co, KGaA, pp. 205–227, published online (2010)
34. Baba, Y., Kurore, R.: Analysis and flamelet modeling for spray combustion. J. Fluid Mech. **612**, 45–79 (2008)
35. Domingo, P., Vervisch, L., Reveillon, J.: DNS analysis of partially premixed combustion in spray and gaseous turbulent flame-bases stabilized in hot air. Combust. Flame **140**, 172–195 (2005)
36. Tanaka, T., Eaton, J.K.: Classification of turbulence modification by dispersed spheres using a novel dimensionless number. Phys. Rev. Lett. **101**, 114502 (2008)
37. Beishuizen, N.A., Naud, B., Roekaerts, D.J.E.M.: Evaluation of a modified Reynolds stress model for turbulent dispersed two-phase flows including two-way coupling. Flow Turbul. Combust. **79**(3), 321–341 (2007)
38. Laín, S., Sommerfeld, M.: Turbulence modulation in dispersed two-phase flow laden with solids from a Lagrangian perspective. Int. J. Heat Fluid Flow **24**, 616–625 (2003)
39. Sadiki, A.: Extended thermodynamics as modeling tool of turbulence fluid flows. Trends in Applications of Mathematics to Mechanics, pp. 451–462, Shaker Verlag, Aachen (2005)
40. Groh, B., Sadiki, A., Janicka, J.: Large eddy simulation of particle-laden vertical channel flow under consideration of a consistent modeling of turbulence modulation. In: Proceedings of the 3rd International Symposium on Two-Phase Flow Modeling and Experimentation, Pisa, Italy (2004)
41. Walklate, P.J.: A Markov-chain particle dispersion model based on air flow data: extension to large water droplets. Bound-Lay. Meteorol. **37**, 313–318 (1986)

42. Hanjalic, K.: Will RANS survive LES: a view of perspectives. ASME J. Fluids Eng. **127**, 831–839 (2005)
43. Wegner, B., Maltsev, A., Schneider, C., Sadiki, A., Dreizler, A., Janicka, J.: Assessment of unsteady RANS in predicting swirl flow instability based on LES and experiments. Int. J. Heat Fluid Flow **25**(3), 528–536 (2004)
44. Janicka, J., Sadiki, A.: Large eddy simulation of turbulent combustion systems. Proc. Combust. Inst. **30**, 537–547 (2005)
45. Poinsot, T., Veynante, D.: Theoretical and Numerical Combustion. R.T. Edwards, Philadelphia, ISBN 1-30217-05-6
46. Kohnen, G., Rüger, M., Sommerfeld, M.: Convergence behavior for numerical calculations by the Euler/Lagrange method for strongly coupled phases. FED-Vol185, Numerical Methods in Multiphases Flows. ASME, Houston Texas (1994)
47. Peters, N.: Fifteen lectures on laminar and turbulent combustion, ERCOFTAC Summer School, p. 174 (1992)
48. Bray, K.N.C., Libby, P.A: Recent developments in the BML model of premixed turbulent combustion. In: Libby, P.A., Williams, F.A. (eds.) Turbulent Reacting Flows, pp. 115–151. Academic, London/San Diego/New York (1994)
49. Gouldin, F.C., Bray, K.N.C., Chen, J.-Y.: Chemical closure model for fractal flamelets. Combust. Flame **77**, 241–259 (1989)
50. Lindstedt, R.P., Váos, E.M.: Modeling of premixed turbulent flames with second moment methods. Combust. Flame **116**, 461–485 (1999)
51. Maltsev, A.: Towards the development and assessment of complete CFD models for the simulation of stationary gas turbine combustin processes. Ph.D. thesis, TU-Darmstadt (2003)
52. Maltsev, A., Sadiki, A., Janicka, J.: A new BML-based modeling for the description of gas turbine typical combustion process. Prog. Comput. Fluid Dyn. **4**(3–4), 229–236 (2004)
53. Muppala, S.P.R., Dinkelacker, F.: Numerical modelling of the pressure dependent reaction source term for premixed turbulent methane/air flames. Prog. Comput. Fluid Dyn. **4**, 328–336 (2004)
54. Le Clercq, P., Doue, N., Rachner, M., Aigner, M.: Validation of a multicomponent-fuel model for spray computations. In: AIAA Aerospace Science Meeting, Orlando, 5–8 Jan 2009
55. Melheim, J., Horender, S., Sommerfeld, M.: Modeling of the vortex-structure in a particleladen mixing-layer. In: Proceeding of FEDSM 2005, ASME 2005, Houston, 19–23 June 2005
56. Minier, J.P.: Lagrangian stochastic modeling of turbulent flows. Advances in Turbulence Modeling. Von Karmon Institute for Fluid Dynamics, Belgium (1998)
57. Yakhot, V., Smith, L.M.: The renormalization group the epsilon-expansion and derivation of turbulence models. J. Sci. Comput. **7**, 35 (1992)
58. Crowe, C.T., Troutt, T.R., Chung, J.N.: Numerical models for two-phase turbulent flows. Ann. Rev. Fluid Mech. **28**, 11–43 (1996)
59. Kumar, K., Sung, C.: Laminar flame speeds and extinction limits of preheated n-decane/O_2/N_2 and n-dodecane/O_2/N_2 mixtures. Combust. Flame **151**(1–2), 209–224 (2007)
60. Löffler, M., Pfadler, S., Beyrau, F., Leipertz, A., Dinkelacker, F., Huai, Y., Sadiki, A.: Experimental determination of the sub-grid scale scalar flux in a non-reacting jet, flow. Flow Turbul. Combust. **81**, 1386 (2007)
61. Gulati, A., Driscoll, J.F. (1986) Velocity-density correlations and Favre averages measured in a premixed turbulent flame. Combust. Sci. Technol. **48**, 285–307 (1986)
62. Videto, B.D., Santavicca, A.: Flame-turbulence interactions in a freely-propagating premixed flame. Combust. Sci. Technol. **70**, 47–73 (1990)
63. Abdel-Gayed, R.G., Al-Khishali, K.J., Bradley, D.: Turbulent burning velocities and flame straining in explosions. Proc. R. Soc. Lond. A **391**, 393–414 (1984)
64. Huang, H., Sobel, D.R., Spadaccini, L.J.: AIAA 2002-3871
65. Zhou, P., Hollis, O.L., Crynes, B.L.: Ind. Eng. Chem. Res. **26** 852 (1987)
66. Honnet, S., Seshadri, K., Niemann, U., Peters, N.: A surrogate fuel for kerosene. Proc. Combust. Inst. **32**(1), 485–492 (2009)
67. Subramanian, G., Colin, O., Pires da cruz, O., Vervisch, L., Brunneaux, G.: Modelling turbulent mixing of an evaporatine spray. In: 20th International Colloquium on the Dynamics of Explosions and Reactive systems. Report No. 138, Montreal, Canada, 31 July – 5 August (2005)

Chapter 4
Toward the Impact of Fuel Evaporation-Combustion Interaction on Spray Combustion in Gas Turbine Combustion Chambers. Part II: Influence of High Combustion Temperature on Spray Droplet Evaporation

Amsini Sadiki, W. Ahmadi, and Mouldi Chrigui

Abstract While the first part of this work dealt with the question related to the impact of partial vaporization on spray flame properties, the second part attempts to answer the question how the evaporation process behaves under turbulent premixed combustion conditions. According to the Eulerian-Lagrangian RANS-based spray module designed in Part I the non-equilibrium evaporation model, that was proven to capture accurately the evaporation process characteristics under operating environments similar to industrial application conditions, is especially applied here. The so-called conditioned progress variable approach (CPVA) based on the Bray-Moss-Libby (BML) model as adapted in Part I to account for both premixed and partially premixed combustion is used to describe the combustion process and determine the gas properties that are seen by droplets. To assess the numerical approach and to analyze the evaporation-combustion interactions a model gas turbine combustor fueled by N-heptane is considered as to exhibit the behaviour of single component fuel. It features a turbulent premixed combustion. To isolate the impact of high combustion temperature on spray evaporation, two cases were investigated

A. Sadiki (✉)
Department of Mechanical and Processing Engineering, Institute for Energy and Powerplant Technology, Technische Universität Darmstadt, Petersenstr. 30, 64287 Darmstadt, Germany
and
Center of Smart Interface, Technische Universität Darmstadt, Petersenstr. 32, 64287 Darmstadt, Germany
e-mail: sadiki@ekt.tu-darmstadt.de

W. Ahmadi
Department of Mechanical and Processing Engineering, Institute for Energy and Powerplant Technology, Technische Universität Darmstadt, Petersenstr. 30, 64287 Darmstadt, Germany
e-mail: ahmadi@ekt.tu-darmstadt.de

M. Chrigui
Department of Mechanical and Processing Engineering, Institute of Energy and Power Plant Technology, Darmstadt University of Technology, Petersenstr. 30, D-64287 Darmstadt, Germany
e-mail: mchrigui@ekt.tu-darmstadt.de

B. Merci et al. (eds.), *Experiments and Numerical Simulations of Diluted Spray Turbulent Combustion*, ERCOFTAC Series 17, DOI 10.1007/978-94-007-1409-0_4,

and compared. In the first case the spray evaporation is studied in a non-combusting environment. In the second it is investigated how the premixed combustion conditions affect droplet spray characteristics.

Keywords N-heptane spray · Evaporation · Premixed flame · Euler-Lagrange · RANS

1 Introduction

As already pointed out in the first part of this work (see Chap. 3), spray behavior has been studied extensively in recent years. Thereby, the quality of atomization along with fuel-air mixture preparation has been recognized as one of the main factors controlling the combustion of liquid fuel sprays. Much attention has been given to methods describing or measuring droplet sizes and distribution as well as evaporation rates of droplets and then to retrieve their effect on spray combustion. Most of the studies focus essentially on flame structure and gas phase properties [1–14]. Contributions devoted to the effects of flame and resulting high temperature conditions, environmental pressure and high-density gradient on the spray behaviour itself are very few. McDonell et al. [1] investigated, experimentally, reacting and non-reacting swirled assisted spray where measurements of drop size and three components velocity were conducted. They also studied in [2] the behaviour of methanol spray produced by atomizer with different operating modes. Detailed measurements were provided and physically interpreted. Marley et al. [3] used laser-induced fluorescence and smoke visualization to investigate a turbulent ethanol spray flame. They generated a polydisperse spray that yielded larger droplets that are able to cross the inner reaction zone and then vaporize in the hot region bounded by the double flame structure. Karpetis and Gomez [4] conducted experimental research on well-defined, turbulent non-premixed spray flames of methanol. They measured velocity and temperature of the gaseous phase, as well as size, velocity, and concentration of the droplet phase, along with the relevant statistical quantities using phase Doppler anemometry and spontaneous Raman thermometry. Stepowski et al. [5] and Bazile and Stepowski [6] focused on the spray flame stabilization. They conducted several researches toward phenomena in gas turbine combustion where fuel is injected as spray. Wildmann and Presser [7] provided an interesting database for multiphase modeling where methanol spray flame was experimentally investigated. Presser et al. [8] examined experimentally, using laser velocimetry and phase Doppler interferometry, droplet size and velocity distribution in a swirl stabilized pressure atomized spray. In [9] they investigated the droplet transport in a swirl-stabilized spray flame.

With focus on N-heptane fuel some DNS investigations are reported by Luo et al. [10] to gain a better understanding of three-dimensional spray flame behaviour, as well as evaporation-combustion interaction in a model swirling configuration scaled down significantly to allow for DNS of all scales. Thereby an unsteady flamelet model was developed to account for the influence of evaporation on scalar dissipation rates. Among others the issues related to flamelet equations for spray combustion considering the influence of droplet evaporation was not addressed. Domingo et al. [11] carried out DNS of partially premixed combustion in vaporizing spray and proposed closures for source terms appearing in the transport equation for the reaction progress variable. Hollmann and Gutheil [12] developed a flamelet model for turbulent spray diffusion flames and incorporated a laminar spray flame library. Applications to methanol/air and to N-heptane have been reported. A recent review of modeling and simulation of spray combustion is reported by Gutheil [13]. Wright et al. [14] carried out experimental and simulations of N-heptane spray auto-ignition at high pressure and high temperature constant volume combustion chamber. For the simulations, the conditional moment closure method has been applied. Yu et al. [15] studied numerically spray combustion characteristics of N-heptane in IC engine configurations. The study employs the hybrid breakup model, high pressure evaporation model and transient flamelet model with detailed chemistry. The results in terms of spray dynamics, evaporation, turbulent mixing, auto-ignition, flame structure, turbulence-chemistry interaction and pollutant (soot, NOx) formation have been provided. They used these data as basis for comparison to dimethyl ether spray combustion in diesel engines. With regard to gas turbine applications, Hage et al. [16], among others, investigated experimentally the flow fields and droplet diameter distributions of water and N-heptane sprays at varied boundary conditions in a generic gas turbine combustor. Chrigui et al. [17–19] used a RANS-based approach similar to that presented in Part I of this work to simulate these configurations. The investigations focused essentially on spray evaporation and dispersion within the carrier phase. The results for a non-reacting configuration [17, 18] and those for a reacting case [19] have been reported, where the effect of pressure variation has been pointed out.

The present work focuses especially on understanding the behavior of the spray characteristics within a turbulent premixed flame, i.e. studying the spray evaporation under high combustion temperature conditions. To assess the numerical approach and analyze the processes under study the model gas turbine combustor fueled by N-heptane used in [18,19] is considered as to exhibit the behaviour of single component fuel. It features a turbulent premixed combustion. In [18,19] the swirler was not included in the computational domain. Rather, appropriate boundary conditions at the inlets were imposed using velocity profiles provided by measurements at 1 mm downstream for the gas phase and at 3 mm for the dispersed phase. In contrast to this, we include in the present study the swirler in the computational domain to improve the inlet conditions description for the simulations. To isolate the impact of combustion on spray evaporation, two cases were investigated and compared. In the first case the spray evaporation is studied in a non-combusting environment. How premixed combustion conditions affect the droplet spray characteristics is investigated in the second case.

The contribution is organized as follows: Sect. 2 recalls the main modelling and numerical procedures. In Sect. 3 the investigated configuration is described along with the boundary conditions used. In Sect. 4 detailed comparisons between numerical results with experimental data in both configurations are provided and discussed in terms of spray dynamics and spray evaporation characteristics. Section 5 is devoted to conclusions.

2 Outline of Combustion Modelling for N-heptane Fuel and Numerical Procedures

As this contribution is concerned with the second part of the work presented in Chap. 3 it will thus be assumed that the reader is familiar with the content of Part I; so we do not recall all the mathematical preliminaries. Nevertheless, to study the interaction between evaporation and combustion along with the essential modeling issues, accurate description of both droplet evaporation and combustion processes is relevant. For this purpose the non-equilibrium evaporation model that has been found to be appropriate for droplet sizes and operating conditions similar to those involved in this configuration, is considered for the description of evaporation process. With respect to the interaction between microprocesses within the spray flow, a two-way coupling is accounted for as a global volume fraction for the two-phase flows equal to 1.09×10^{-4} has been measured experimentally. The turbulence modulation model by Sadiki et al. (see Part I) is applied. The so-called conditioned progress variable approach (CPVA) based on the Bray-Moss-Libby (BML) model as adapted in Part I to account for both premixed and partially premixed combustion is used to describe the premixed combustion process and determine the gas properties that are seen by droplets. Only specific aspects of this combustion treatment for N-heptane fuel are outlined here. In particular the parameter s_L^0 in Eqs. 3.23 and 3.25 that denotes the laminar burning velocity is approximated by Müller et al. [20] for the lean N-heptane-air mixture in the range of preheat temperature between 298 and 800 K and pressure between 1 and 40 bar. The equation of the laminar burning velocity can be derived from the asymptotic analysis carried out by Peters and Williams [21] as:

$$s_L^0 = A\left(T^0\right) Y_{F,u}^m \frac{T_u}{T^0} \left(\frac{T_b - T^0}{T_b - T_u}\right)^n, \tag{4.1}$$

where the function $A(T^0) = F \exp\left(-\frac{G}{T^0}\right)$ depends on thermodynamic and kinetic properties. In (4.1) T^0 is the inner layer temperature and depends on fuel. For N-heptane, T_0 is set to 1,250 K [20]. $Y_{F,u}^m$ denotes the mass fraction of the fuel in the unburned gas. The variable exponents *m, n*, and the equation constants *G* and *F* are given in Table 4.1 for N-heptane [20].

Table 4.1 Approximation constants for burning velocity [20]

Fuel	F [cm/s]	G [K]	m	n
$n\text{-}C_7H_{16}$	7.956×10^3	912.00	0.52	2.30

Note that Eqs. 4.1 and 3.24 together in (3.23) lead to an expression similar to that derived by Metghalchi and Keck [22] or Dahoe and de Goey [23] expressing the dependency of the laminar burning velocity on the temperature and pressure of the mixture as a polynomial function.

As the computational method is based on an Euler-Lagrangian coupling approach, two solvers have been involved and coupled. A 3-D Lagrangian code (LAG3D) for droplet tracking and the three dimensional CFD-code FASTEST in which the carrier phase equations are solved as described in Part I. How these two solvers are numerically coupled has been already outlined in Part I, Sect. 2.5. To note is that 400,000 droplets were tracked every coupling. Due to the low droplet loading only four couplings were sufficient to reach acceptable convergence.

3 Configurations and Boundary Conditions

3.1 Geometry

The simulated complex geometry resembles the configuration experimentally investigated by Hage et al. [16] as depicted in Fig. 4.1. It is composed of a combustion chamber supplied with an air blast atomizer (Fig. 4.2), with an injection nozzle that

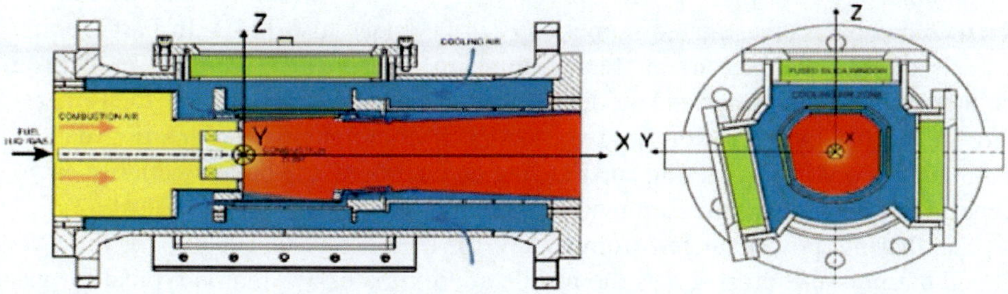

Fig. 4.1 Longitudinal cross section of the combustion chamber as it is designed experimentally [16–19]

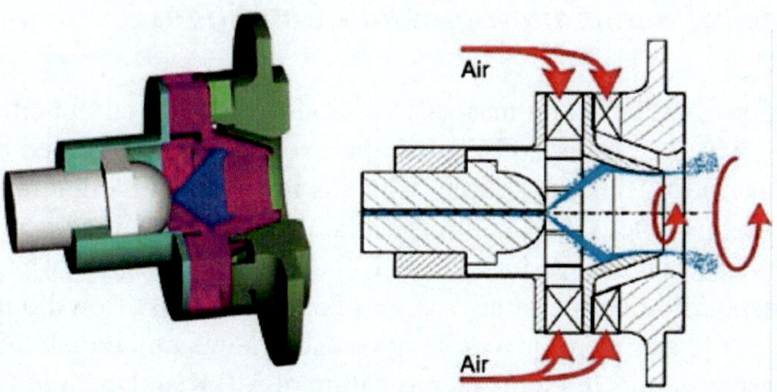

Fig. 4.2 Schematic and 3-D view of the air-blast atomizer [16–19]

Table 4.2 Summary of operating conditions [16]

Case	P [bar]	Air temperature [K]	Air mass flux [g/s]	Equivalence ratio [–]	Re_{air} [10^3]
A	2	523	24.9	0.9	35.2
B	2	523	27.2	0.9	38.4
C	2	623	24.9	0.9	31.2
D	3	623	37.4	0.9	46.8

Re_{air} is calculated based on hydraulic and bulk velocities

spreads the fuel onto a conical sheet. The fuel is then exposed to the high-velocity swirled airstreams of heated carrier phase having high turbulence intensity. This ensures a reliable mixing efficiency.

As cooling is relevant, cold air flows on the outer side of the combustor and penetrates into the second half of the combustion chamber. To characterize the gas phase variables, Laser Doppler Anemometry (LDA) was used in combination with an argon laser. A two-component Phase Doppler Anemometry (PDA) was used to simultaneously measure droplet velocities. The measurements were taken at different flow conditions for the airflow rate, droplet velocities, droplet diameter distribution and liquid flow rate to provide a set of reliable data. The measurements were taken in various cross-sections downstream the inlet for droplets and carrier phase providing a valuable database for model validation of polydispersed sprays.

Various cases that were experimentally investigated in [16] using N-heptane fuel are summarized in Table 4.2. The influence of pressure variations on spray characteristics has been investigated numerically in [19] (see also [24]). In Part I, some aspects of the influence of air temperature on the spray combustion have been pointed out in a pre-vaporized model combustor fired by kerosene. We focus therefore in the present study on case C for which comprehensive validation data is available for both non-reacting and reacting cases. For all the details with respect to experimental setup and measurements techniques used, please refer to [16].

The thermal power ranged from 67 kW for the case A to 100 kW for the case D and the pressure drop across the nozzle approximates 3% that is typical for gas turbine applications.

3.2 Numerical Setup and Boundary Conditions

The combustion chamber is numerically represented by the computational domain shown in Fig. 4.3. In contrast to [17–19], the overall mesh contained 800,000 control volumes including the swirl burner which is resolved with eight radial and eight tangential channels where inlet boundary conditions are specified according to experiments. Focused on case C in Table 4.2, the absolute pressure was $P_{inl} = 2$ bar and the corresponding temperature $T_{inl} = 623$ K. The mass flow boundary conditions for the gas phase were provided by measurements and equaled 24.9 g/s. The dispersed phase was injected with a temperature of 313 K and the mass flow equaled 1.49 g/s. The boundary conditions (BC) for the dispersed phase were derived from

Fig. 4.3 Computational model of the swirler nozzle and test rig (not to scale). Detailed view of the tangential swirler channels and the central fuel injection nozzle (*left*). Complete computational domain used (*right*)

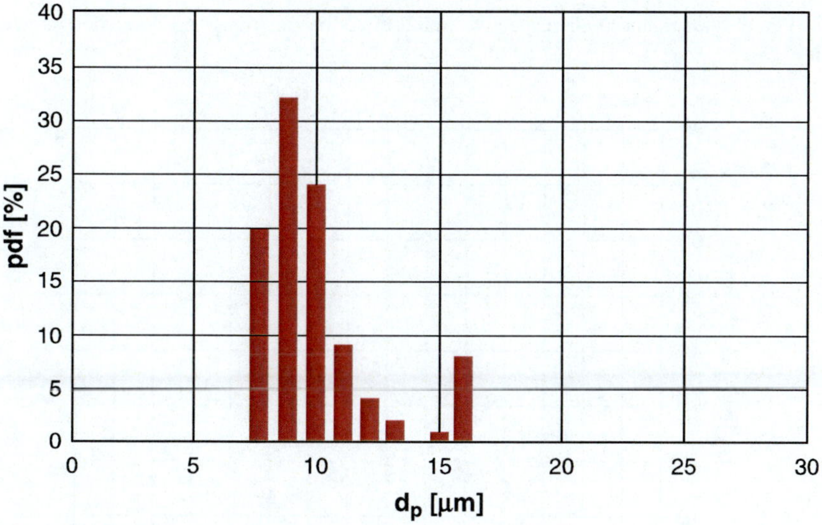

Fig. 4.4 Droplet size distribution at inlet

the experimental data. Since the measurements at 1 mm are associated with high error due to the closeness to the nozzle, the BC for the dispersed phase were adapted to fit to the results at 3 mm. The measured data were performed in eight points radially that define the number of droplet classes. Figures 4.4 and 4.5 show the eight droplet classes. Figure 4.4 represents the initial distribution of the particle size upon which the dispersed mass flux bas been determined, and Fig. 4.5 the corresponding injection position and the relative number of parcels. Each class distinguishes with the others by the parcels number, velocities, rms, diameter and injection locations. The dispersed phase velocities fluctuations at the inlet are displayed in Fig. 4.6. One notices that the established droplet turbulence intensity for the lower classes (droplet diameter up to 12 µm) is about 25% which is similar to that of the carrier phase.

As already mentioned 400,000 parcel trajectories were calculated every coupling iteration to get statistically reliable results. A further increase in the parcel number had no influences on the droplet properties.

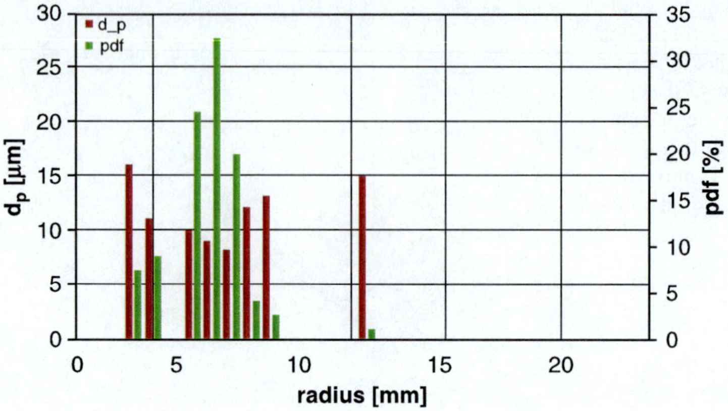

Fig. 4.5 Parcel classes and their respective radial injection positions and corresponding relative quantities

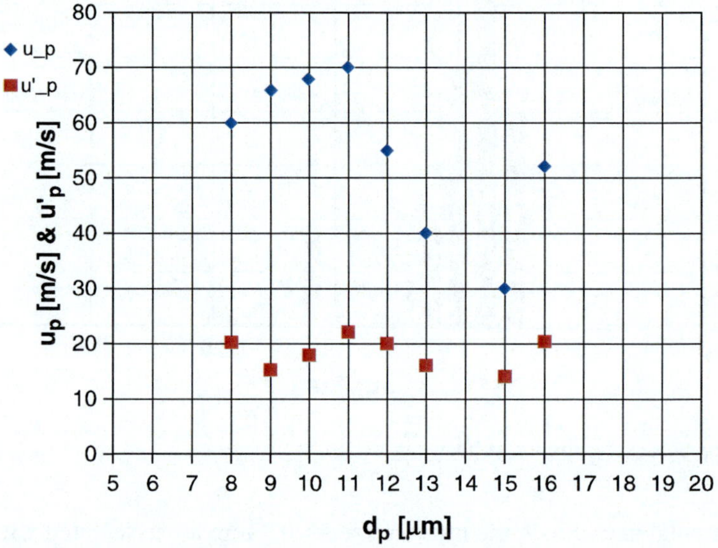

Fig. 4.6 Boundary conditions for the axial droplet velocity and the RMS (m/s)

4 Results and Discussion

4.1 Carrier Gas Properties in Non-Reacting Case

Even though results with respect to carrier gas properties have been already reported in [17] without including the swirler into the computational domain, we achieve here a similar good agreement between experimental data and numerical results for the velocity field and the turbulent kinetic energy. Including the swirler does not

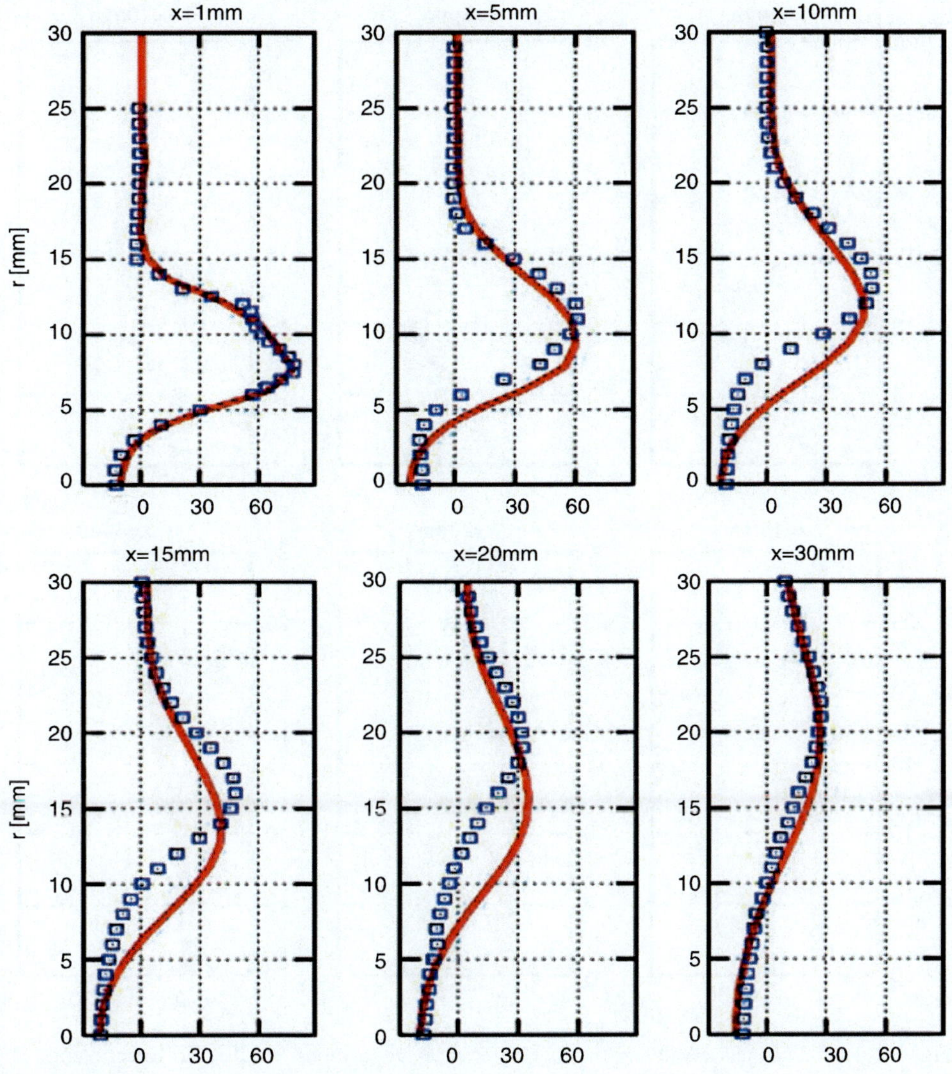

Fig. 4.7 Axial gas phase velocity (m/s) at different axial positions. □□□□□ Experiment; ----- simulation

bring new insights with respect to mean values. The advantage of this inclusion may be seen in capturing unsteady behaviors by the means of URANS as demonstrated in [25]. This task is rather out of the scope of the present work. The results are shown in Figs. 4.7–4.9 for the axial, radial and tangential velocity profiles at axial locations of 1, 5, 10, 15, 20 and 30 mm downstream the inlet. The flow features an increasing radial shift and declining of the velocity peaks with increasing distance from the nozzle. Remarkable deviations between experimental and numerical results are observed for the axial and radial velocity component at x = 10, 15 and 20 mm while the tangential velocity component exhibits evident deviations to experimental data

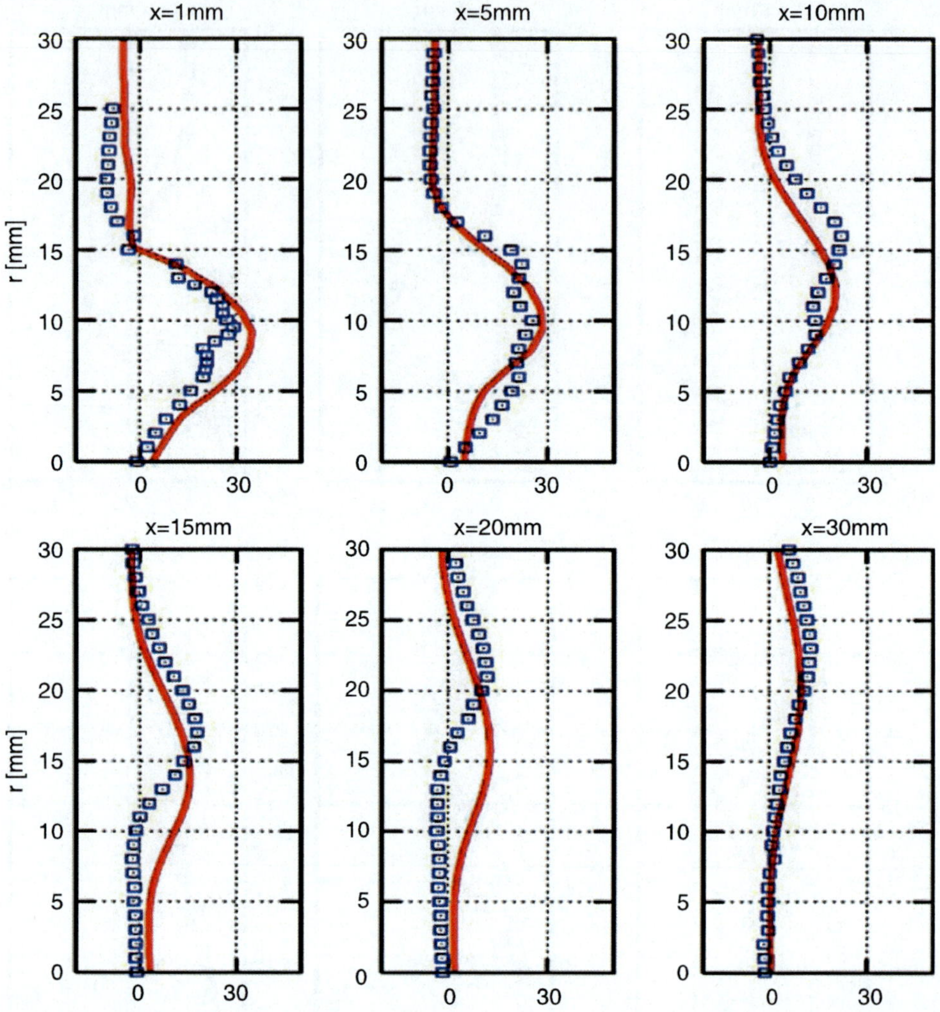

Fig. 4.8 Radial gas phase velocity (m/s) at different axial positions. □□□□□ Experiment; --- simulation

up to x = 30 mm. These are linked to the weakness of the RNG-k-Epsilon turbulence model used. However, the internal recirculation zone (IRZ) is well reproduced by the simulations. Its length is about three times the exit nozzle radius.

It is also of interest to judge the ability of the numerical model to capture the carrier phase flow fluctuations. The turbulent kinetic energy of the continuous phase is depicted in Fig. 4.10. As in experiments, peak values for the turbulent kinetic energy are predicted located in distinct shear layers. A satisfactory agreement can be reported in spite of some remarkable discrepancies between experimental data and numerical predictions within the recirculation zone.

All profiles for the mean velocity components as well as the kinetic energy showed nearly axis-symmetrical behaviour. This indicates that within the investigated region the asymmetric flame tube shape did not significantly influence the

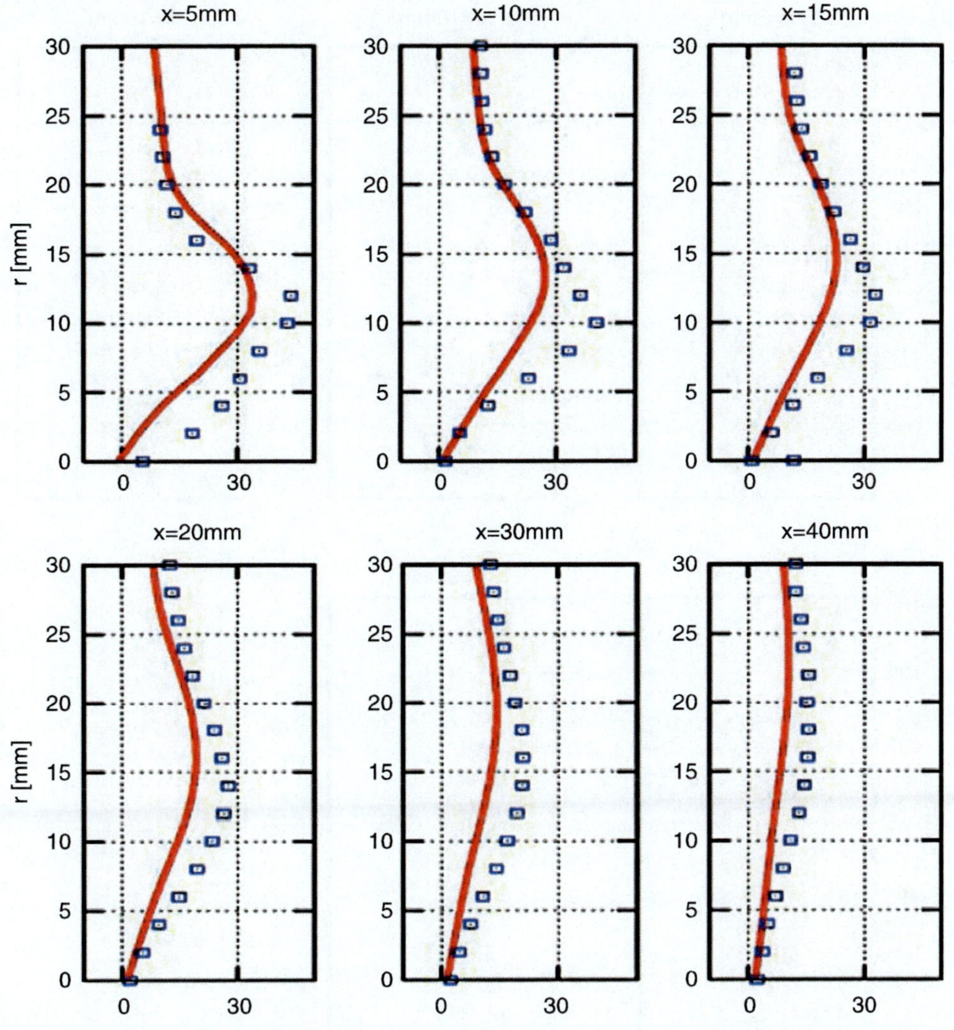

Fig. 4.9 Tangential gas phase velocity (m/s) at different axial positions. □□□□□ Experiment; --- simulation

mean flow field properties. The flow field prediction achieved qualifies the tool for a study of the turbulent droplet dispersion and convection processes around the droplets.

4.2 Turbulent Droplet Dispersion in Non-reacting Case

The capability of the Lagrangian tracking approach to capture the dynamic droplet behavior is evaluated by comparing calculated droplet velocities with experiments.

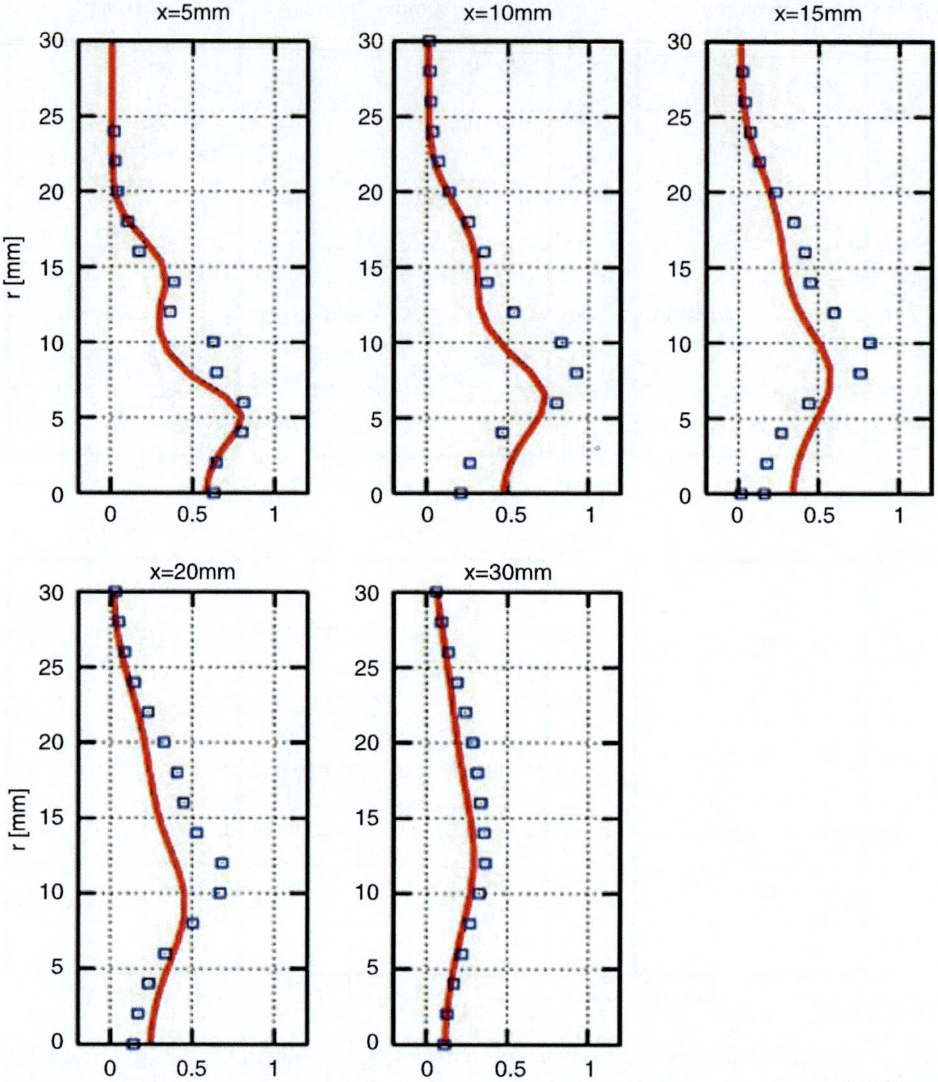

Fig. 4.10 Turbulent kinetic energy of the gas phase (10^3 m^2/s^2) at different axial positions. □□□ Experiment; ___ simulation

Figures 4.11 and 4.12 show the distribution of the dispersed phase axial velocities and their fluctuations. Good agreement between numerical and experimental results for the velocity is observed in Fig. 4.11. The maximum discrepancy between numerical simulations and experimental data is observed at x = 15 mm for the droplet axial velocity and equals 12.6 m/s. This shall be directly linked to that in the carrier phase (Fig. 4.7) as the Stokes number between 0.51 and 4.17 enables the droplets to follow the carrier phase.

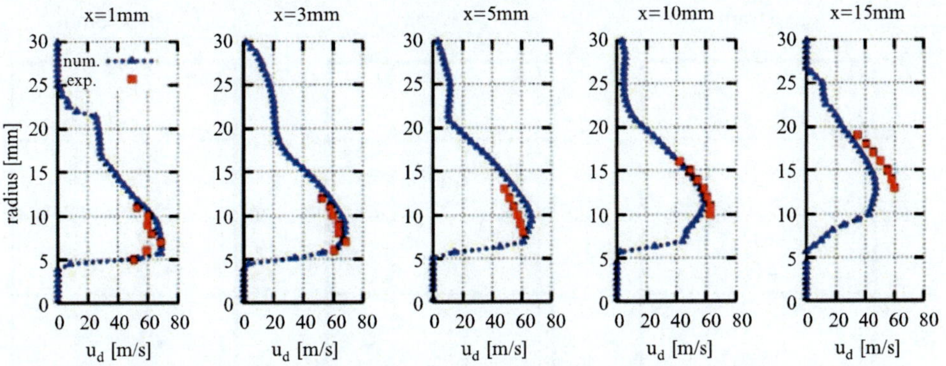

Fig. 4.11 Radial profile of the droplet axial velocity at different axial positions

4.3 *Spray Evaporation in Non-reacting Case*

As expressed in their Eqs. 3.9–3.12, the mass and heat transfer rates are linked to the time evolution of the droplet diameter and temperature. The effect of droplet vaporization can therefore be captured by considering the distribution of droplet diameter along with the droplet mass flux. Thus, the droplet volume flux, which is related to the droplet mass flux by the local density, can be used as a reliable indicator of the evaporation rate validation. Figure 4.13 presents the computed droplet axial volume flux and Fig. 4.14 displays the droplet radial volume flux in comparison to experimental data, respectively. The volume flux in the radial direction is lower than that in the axial one, since droplets are being dragged faster axially. In accordance with experiments the concentration of droplets decreases while moving away from the nozzle due to evaporation. A very good agreement is noticed in all cross sections, except at $x = 5$ mm where the experimental data exhibit an increase of the volume flux. This enhancement in the volume or mass flux cannot be physically explained, since no vapour condensation can occur at the above mentioned operating conditions.

It is observed that the spray width at $x = 3$ mm is about one third of the combustion chamber radius. The spray width is closely correlated with the spray apex angle, which in turn can be modified by the internal recirculation zone (IRZ) of the carrier phase. The droplets are able to survive until the axial position $x = 15$ mm, while they are completely evaporated within the IRZ. One should mention at this stage that the applied temperature boundary condition has a stronger influence on the evaporation rate than the pressure. The experimental measurements along with numerical simulations have acknowledged that a diminution of $\Delta T = 100\,°C$ in the inlet temperature affects the evaporation much more than the reduction of the inlet pressure to 1 bar. Some other works by the authors dealing with other configuration [19, 24] have confirmed this finding as well.

The apex angle of the spray was estimated to 15°, whereas the experiments resulted in 11°. The discrepancy between the measured and calculated values is caused by the disagreement in the prediction of the volume flux (Figs. 4.13 and 4.14). This

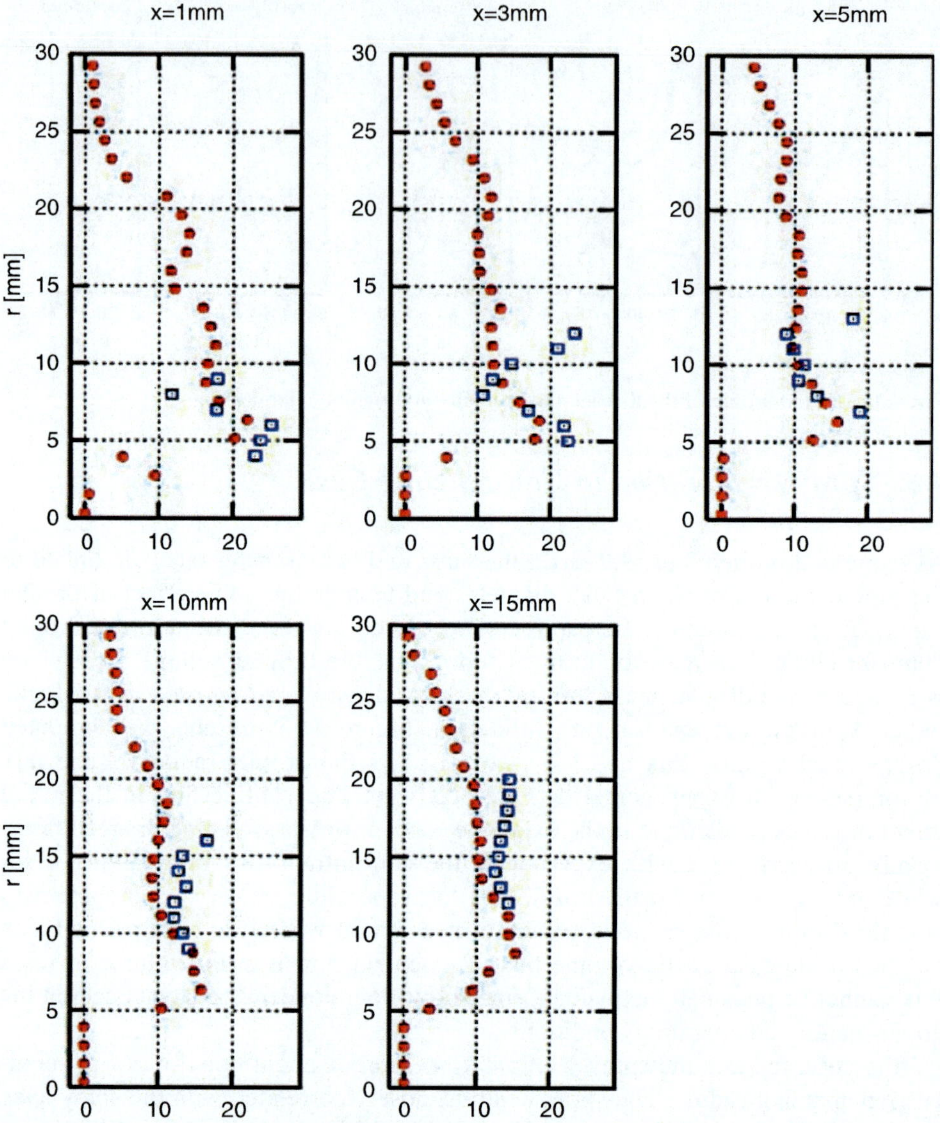

Fig. 4.12 Droplet axial velocity fluctuations (without combustion) (m/s). □□□ Experiment; ●●● simulation

is due to the swirling effect that drags the droplets radially outwards and causes divergent behavior due to centrifugal forces. Indeed one observes clearly, at section 1 mm and also at 3 mm that the predicted droplet volume flux display a broader curves compared to the experimental one, i.e. the spray tends to diverge while moving forward. Note that the volume flux is determined by the droplet axial velocity and the particle size distribution in the cross section which, in turn, is a result of the droplet motion described by its equation including the dispersion. The apex angle

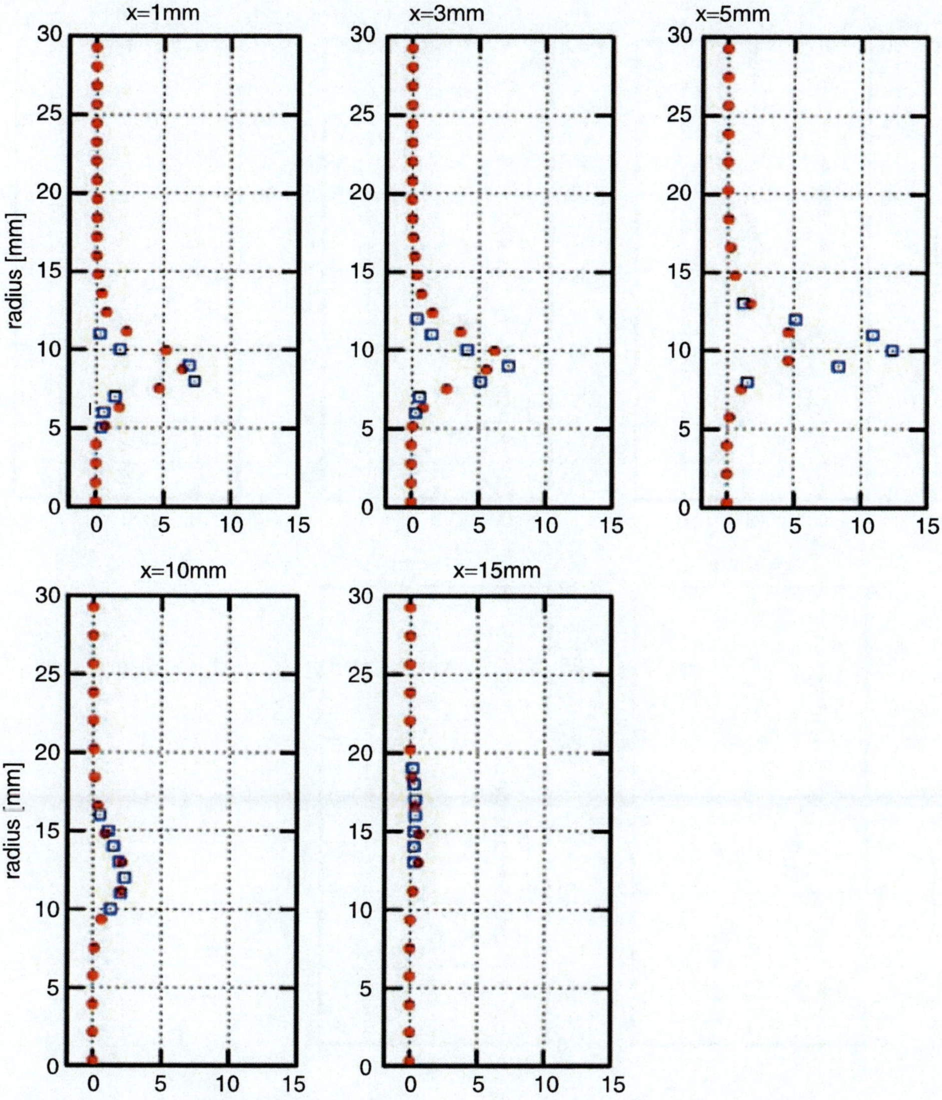

Fig. 4.13 Radial profile of the droplet axial volume flux at different axial positions. □□□ Experiment; • • • simulation

of the spray is therefore influenced by different parameters. It is worth noting that the measurement of the apex spray angle is associated with relatively high error due the dense spray conditions close to the nozzle.

4.4 Turbulent Droplet Dispersion in Reacting Case

Under combustion conditions the N-heptane droplets are subjected to the same operating conditions as before. However, the premixed fuel/air in this case burns while

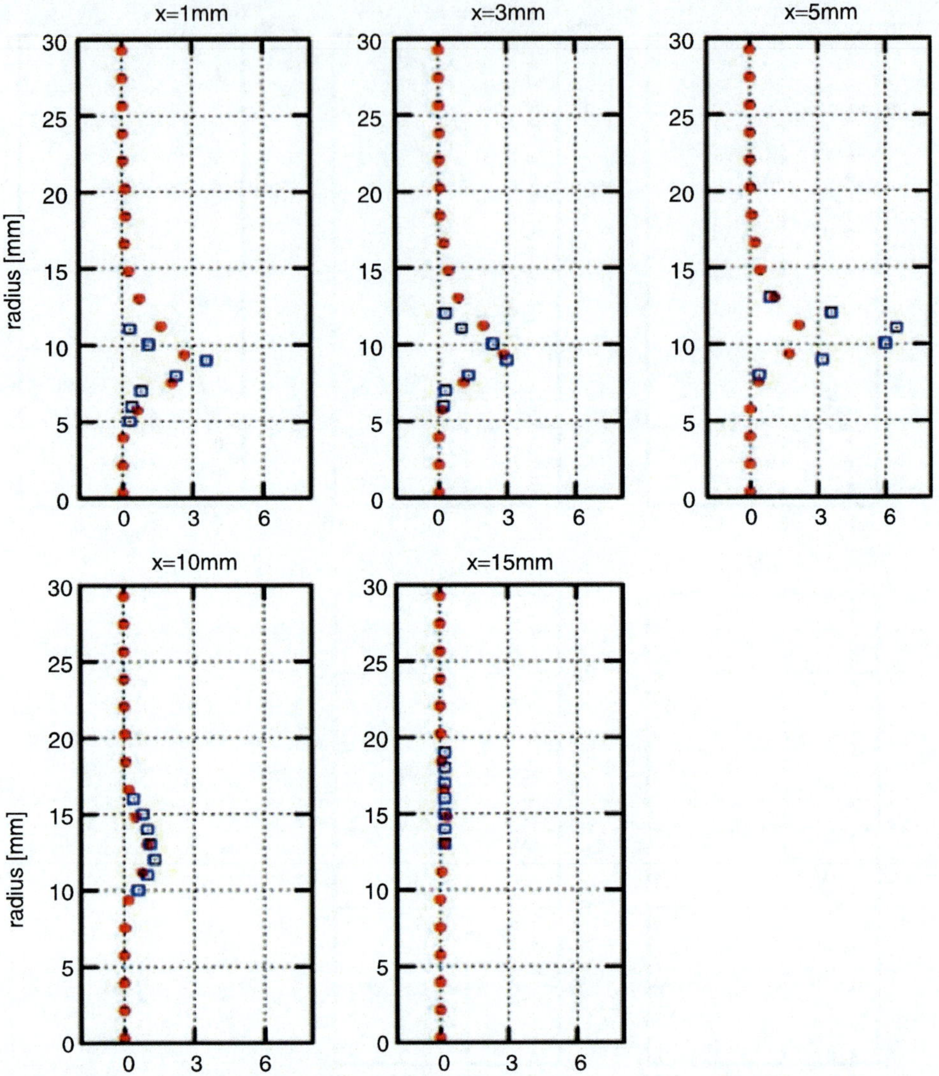

Fig. 4.14 Radial profile of the droplet radial volume flux at different axial positions. □□□ Experiment; • • • simulation

neglecting the droplet-flame interaction. By comparing the spray behaviour in the case with and without combustion, it is possible to gain information about the effect of combustion on N-heptane droplet evaporation. Figures 4.15 and 4.16 display the axial and radial velocities in comparison to experimental data, respectively. A reasonable agreement is observed.

Due to the centrifugal forces, the radial velocity component is increasing while droplets move downstream. The gas phase momentum transfer from the axial to the radial direction during the flow propagation could explain this. At the central axis of

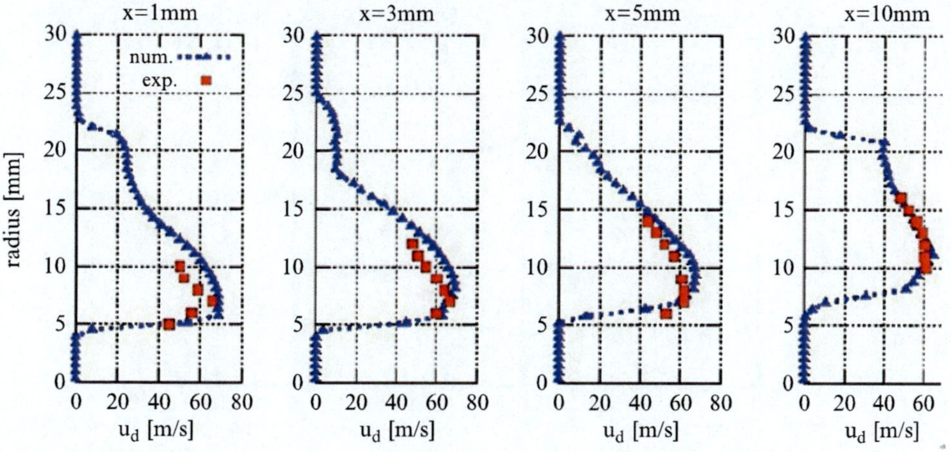

Fig. 4.15 Radial profile of the droplet axial velocity at different axial positions

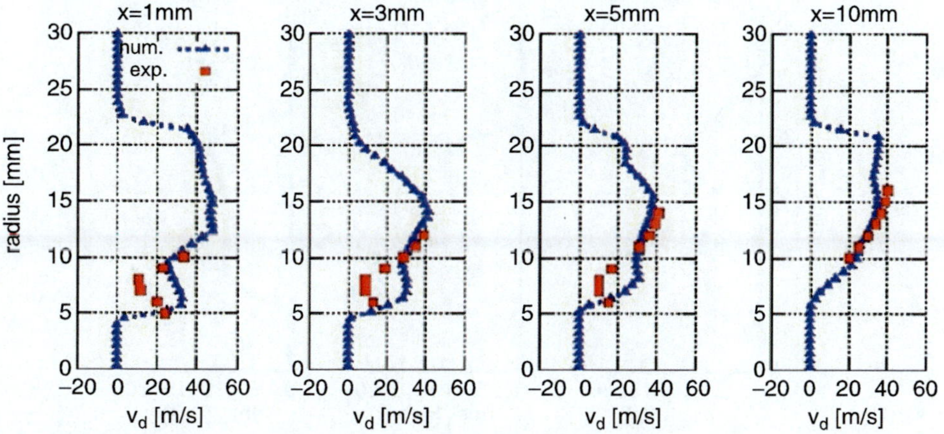

Fig. 4.16 Radial profile of the droplet radial velocity at different axial positions

the combustion chamber, no droplets are observed due to the swirling fluid motion. Therefore, they do not guarantee enough results to provide reliable statistics for this region. It has to be noted that the secondary recirculation zone is small and does not reach the spray jet. For that reason, the fuel will be more confined to the core of the combustion chamber. This has an important influence on the droplet dispersion. In fact, if the droplets are able to reach the secondary recirculation zone, the fuel-air homogeneity will be enhanced and evaporation gets faster due to negligible vapor concentration in this area. Droplets that enter the secondary recirculation zone are not able to leave it. They supply their complete mass to this lean zone.

A higher mixing efficiency and degree of evaporation is therefore obtained. By increasing the radial velocity, the droplets can be transported radially outward more quickly. This helps to disperse droplets into the combustion chamber more

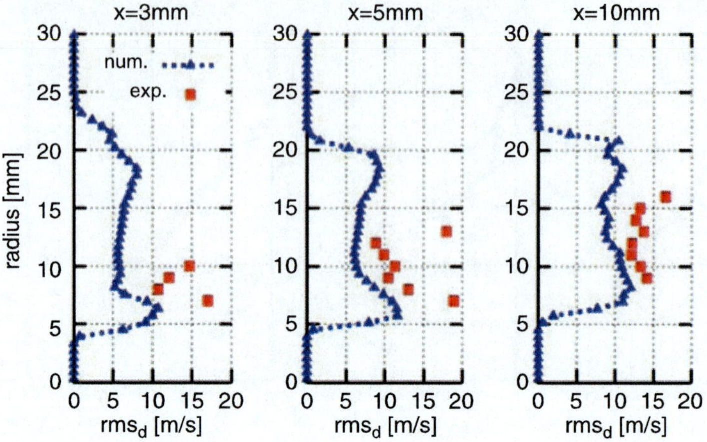

Fig. 4.17 Radial profile of the droplet axial velocity fluctuation (RMS) at different axial positions

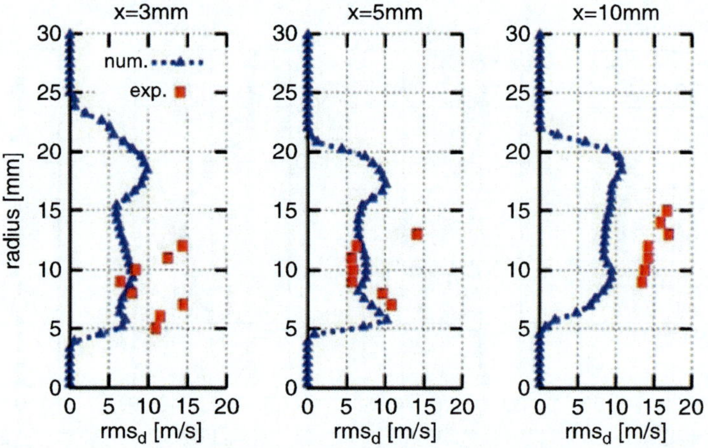

Fig. 4.18 Radial profile of the droplet radial velocity fluctuation (RMS) at different axial positions

efficiently. It is worth mentioning that inflow conditions for the dispersed phase play a major role in the droplet dispersion and, above all, influence the degree of evaporation. Even though primary and secondary break up phenomena have not been considered, appropriate inflow conditions have been used according to experimental data as outlined above (see Figs. 4.4–4.6).

Figures 4.17 and 4.18 present the droplet velocity fluctuations, i.e. root mean square (rms) for the axial and radial directions, respectively. Both RMS profiles do not show considerable turbulence decay while progressing forward. The agreement between measurements and simulations is reasonable. It is noteworthy that the experimental profiles feature peaks at r $\approx$ 7 and 11 mm. These local maximums are related to the high velocity gradients, which are responsible for the production of turbulence. As in Fig. 4.10, this effect could not be captured numerically due essentially to the isotropic nature of the turbulence model.

4.5 Spray Evaporation in Reacting Case

No remarkable differences in the droplet velocities could be detected experimentally or numerically when compared with results in the previous subsection. This originates from the fact that the droplets close to the air-blast atomizer are affected mainly by the boundary conditions.

The droplets travel a short distance no longer than 20 mm, and then vanish. They do not feel the high temperature gradient caused by the flame, because this is established around the fuel concentration as shown in the colored line plots of the temperature in Fig. 4.19. This figure displays the spray concentrations in kilograms and the vector plots of the carrier phase. On the centerline, the droplets are exposed to the warm gas back-flow, which has no effect on the droplet velocities since they are generated starting from the radial position of 5 mm. The N-heptane droplets are actually embedded in similar conditions to the case without combustion, considering the fact that carrier phase characteristics are not changed until 20 mm forwards, which represents the endurance distance for the dispersed phase.

Figure 4.20 presents the droplet Sauter mean diameter at different axial locations. Despite of some discrepancies, a plausible agreement between computational results and experimental data was observed attesting that the evaporation model describes the phase transition under the described operating conditions well.

Figure 4.20 also indicates that smaller droplets are located at the jet edges. This behavior is less pronounced as they move toward the combustion chamber exit. Smaller droplets exhibit higher axial velocity than the larger ones. Similar to non-reacting case the maximum relaxation time is almost 1.2×10^{-4} s whereas the Stokes number varies between 0.51 and 4.17. Thus the smaller droplets are able to follow the carrier phase and accelerate faster compared to the larger droplets.

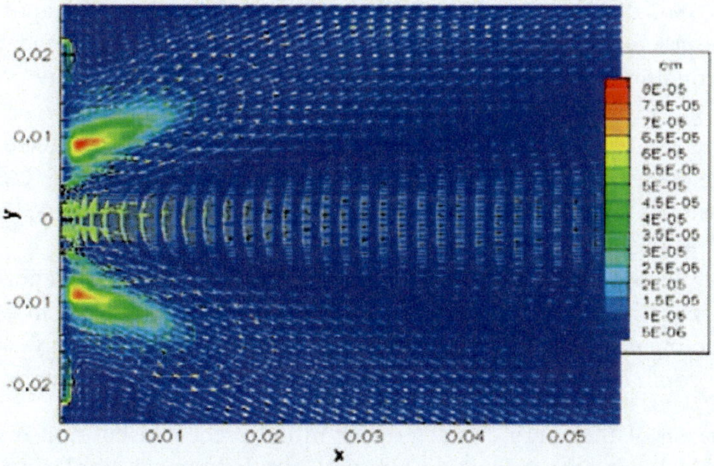

Fig. 4.19 Spray concentration embedded in the carrier phase temperature and velocity vector plots

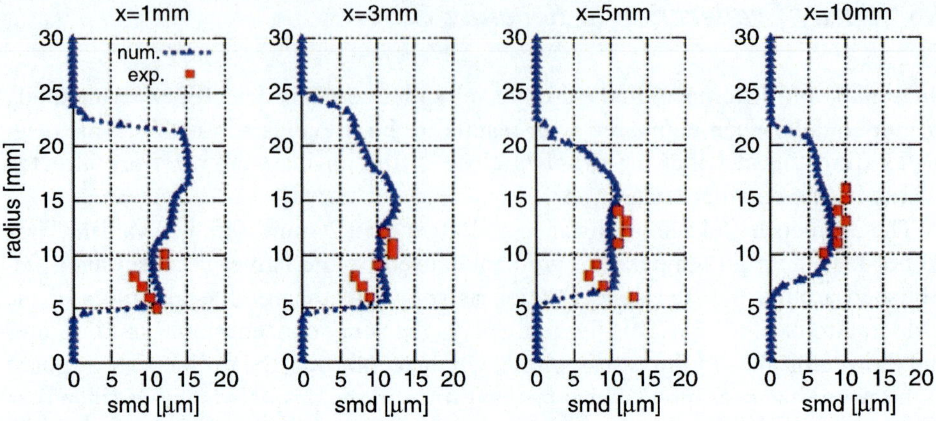

Fig. 4.20 Radial profile of the droplet Sauter mean diameter (SMD) at different axial positions

Equilibrium between the droplets and the carrier phase is observed only after a distance of 15 mm from the configuration inlet, indeed the droplets at this distance have diameters close to that of the tracer particles and perfectly follow the flow paths of the gas phase.

5 Conclusions

This contribution allowed to assess the numerical and modelling approach developed in Part I (Chap. 3) for turbulent spray combustion applications dealing with both partially premixed combustion regime and premixed combustion environments. So, the impact of interaction between fuel droplet evaporation and combustion processes on spray combustion has been first analyzed under partially premixed conditions in Part I. In the present Part II, turbulent spray combustion in a model gas turbine combustor featuring a premixed combustion has been considered. Thereby we investigated how the droplet vaporization is affected by high temperature conditions resulting from premixed combustion using a RANS-based Euler-Lagrangian procedure. No effect of primary break up and coalescence near the nozzle has been considered.

To isolate the impact of high combustion temperature on spray evaporation, two cases were investigated. In the first case, spray evaporation took place in a preheated but non-reacting gas mixture while the second considered droplet spray evaporation affected by combustion conditions. Following concluding remarks can be drawn:

1. Comparison of the spray characteristics in the preheated but non-reacting gas mixture to those with a reacting mixture revealed almost similar results. This originates from the fact that N-heptane droplets undergo similar conditions. The

differences between reacting and non-reacting conditions could be observed only 20 mm from the inlet downstream, which could not be reached by the spray.
2. A satisfactory agreement between experimental data and numerical results has been achieved. However, most discrepancies observed could be clearly linked to the weakness of the turbulence model used. This makes clear why advanced turbulence models have to be applied for future investigations.
3. In particular the need of turbulent structures-resolving models, like Large Eddy Simulation (LES), are highly demanded to accurately capture unsteady behavior of the carrier phase flow and the spray dynamics. This shall also take better advantage of the inclusion of the swirler nozzle into the computational domain.
4. By using numerical simulations it is possible to assess the validity of experimental data. Curious deviations between experimental data and simulation results observed for the droplet volume flux at $x = 5$ mm revealed possible errors in measurements data exploitation.
5. The designed spray module has proven to reliably predict flow and evaporating droplet properties under industrial gas turbine conditions in which droplet-flame interaction is negligible. In that case the combustion models of gaseous combustion have been used as first approximations. The direct influence of droplet evaporation source in flamelet-based modelings needs to be accounted for and requires further work that is left for future.

As pointed out above unsteady calculation techniques like URANS and LES will be most useful. For both URANS and LES, the precision of results is often evaluated by a comparison with experimental data. A systematic LES-model assessment should actually include the computations and comparison of the numerical results to experimental data in configurations of various complexities. Distinct aspects have to be investigated, first separately and thereafter collectively, increasing the complexity of the investigated system. Endeavours directed to include these aspects as pointed out during the first "workshop on turbulent spray combustion" must be pursued.

Acknowledgment The authors acknowledge the financial support from the German Research Council (DFG) through the SFB568 and GRK1144.

References

1. McDonell, V.G., Adachi, M., Samuelsen, G.S.: Structure of reacting and nonreacting, non-swirling, air-assisted sprays, part I: gas phase properties. Atomization Sprays **3**(4), 389–410 (1993)
2. McDonell, V.G., Samuelsen, G.S.: An experimental data base for the computational fluid dynamics of reacting and non-reacting methanol sprays. J. Fluids Eng. **117**, 145–153 (1995)
3. Marley, K., Welle, E.J., Lyons, K.M., Roberts, W.L.: Effects of leading edge entrainment on the double flame structure in lifted ethanol spray flames. Exp. Therm. Fluid Sci. **29**(1), 23–31 (2004)
4. Karpetis, A.N., Gomez, A.: An experimental investigation of non-premixed turbulent spray flames and their self-similar behavior. Proc. Combust. Inst. **27**(2), 2001–2008 (1998)

5. Stepowski, D., Cessou, A., Goix, P.: Flame stabilization and OH fluorescence mapping of the combustion structures in the near field of a spray jet. Combust. Flame **99**, 516–522 (1994)
6. Bazile, R., Stepowski, D.: Measurements of vaporized and liquid fuel concentration fields in a burning spray jet of acetone using planar laser induced fluorescence. Exp. Fluids **20**, 1–9 (1995)
7. Wildmann, J.F., Presser, C.: A benchmark experimental database for multiphase combustion model input and validation. Combust. Flame **129**, 47–86 (2002)
8. Presser, C., Gupta, A.K., Semerjian, H.G.: Aerodynamic characteristics of swirling spray flames: pressure-jet atomizer. Combust. Flame **92**, 25–44 (1993)
9. Presser, C., Gupta, A.K., Avedisian, C.T., Semerjian, H.G.: Droplet transport in a swirl-stabilized spray flame. J. Propul. Power **10**, 631–638 (1994)
10. Luo, K., Pitsch, H., Pai, M.G.: DNS of three-dimensional swirling n-heptane spray flames. Center for Turbulence Research, Annual Research Briefs, pp. 171–183 (2009)
11. Domingo, P., Vervisch, L., Reveillon, J.: DNS analysis of partially premixed combustion in spray and gaseous turbulent flame-bases stabilized in hot air. Combust. Flame **140**, 172–195 (2005)
12. Hollmann, C., Gutheil, E.: Flamelet-modelling of turbulent spray diffusion flames based on a laminar spray flame library. Combust. Sci. Technol. **135**, 175–192 (1998)
13. Gutheil, E.: Modeling and simulation of droplet and spray combustion. Handbook of Combustion, vol. 5. Wiley-VCH Verlag GmbH & Co, pp. 205–227, published online (2010)
14. Wright, Y., Nektaria, O.M., Konstantinos, B., Giorgio, P., Espaminondas, M.: Experiments and simulations of n-heptane spray auto-ignition in a closed combustion chamber at diesel engine conditions. Flow Turbul. Combust. **84**(1), 49–78 (2009)
15. Yu, Y., Kang, S., Kim, Y., Lee, K.S.: Numerical study on spray combustion processes in n-heptane and dimethyl ether fueled diesel engines. Energy Fuels **23**(10), 4917–4930 (2009)
16. Hage, M., Dreizler, A., Janicka, J.: Flow fields and droplet diameter distributions of water and n-heptane sprays at varied boundary conditions in a generic gas turbine combustor. ASME-GT-2007-27108, Montreal, Canada (2007)
17. Chrigui, M.: N-heptane spray evaporation and dispersion in turbulent flow within a complex geometry configuration. J. Comput. Therm. Sci. **2**(1), 55–78 (2010)
18. Chrigui, M., Hage, M., Sadiki, A., Janicka, J., Dreizler, A.: Experimental and numerical analysis of spray dispersion and evaporation in a combustion chamber. Atomization Spray **19**, 929–955 (2009)
19. Chrigui, M., Zghal, A., Sadiki, A., Janicka, J.: Evaporation and dispersion of n-heptane droplets within premixed flame. J. Heat Mass Transfer **46**(8–9), 869–880 (2010)
20. Müller, U.C., Bollig, M., Peters, N.: Approximations for burning velocities and Markstein numbers for lean hydrocarbon and methanol flames. Combust. Flame **108**, 349–356 (1997)
21. Peters, N., Williams, F.A.: The asymptotic structure of stoichiometric methane-air flames. Combust. Flame **68**, 185–207 (1987)
22. Metghalchi, M., Keck, J.C.: Burning velocities of mixtures of air with methanol, iso-octane, and indolene at high pressure and temperature. Combust. Flame **48**, 191–210 (1982)
23. Dahoe, A.E., de Goey, L.P.H.: On the determination of the laminar burning velocity from closed vessel gas explosions. J. Loss Prev. Process Ind. **16**, 457–478 (2003)
24. Chrigui, M., Roisman, I., Batarseh, F.Z., Sadiki, A., Tropea, C.: Spray generated by an airblast atomizer under elevated ambient pressures. J. Propul. Power **26**, 1170–1183 (2010)
25. Wegner, B., Maltsev, A., Schneider, C., Sadiki, A., Dreizler, A., Janicka, J.: Assessment of unsteady RANS in predicting swirl flow instability based on LES and experiments. Int. J. Flows Fluid Flow **25**, 528–536 (2004)

Chapter 5
Reynolds Stress and PDF Modeling of Two-Way Coupling and Vaporisation Interaction in a Turbulent Spray Flame

Nijso A. Beishuizen and Dirk Roekaerts

Abstract In turbulent spray flames the momentum and mass transfer between fuel droplets and gas phase influence the turbulence properties of the gas phase. The importance of these two-way coupling effects are investigated numerically by means of a transported Probability Density Function (PDF) method and compared with experimental data of a turbulent non-premixed methanol spray flame obtained by Karpetis and Gomez. We focus on the development of a Reynolds stress model that takes into account the presence of particles and vaporisation. A hybrid Finite Volume/ Monte Carlo method is used to solve the gas phase properties and a Lagrangian particle method is used to solve the dispersed phase properties. The model for the pressure-rate-of-strain is extended to take into account interaction effects and an exact model for the effect of vaporisation on the turbulence dissipation is proposed. The drag reduction and the heat and mass transfer coefficient of the fuel droplets due to vaporisation is taken into account based on numerical correlations found by Chiang, Raju and Sirignano. Comparison between simulations and experimental data show that two-way coupling and vaporisation prevent the droplets from completely relaxing to the gas phase conditions. The two-way coupling terms appearing in the model for the pressure-rate-of-strain are of the same order of magnitude as the classical two-way coupling terms and can therefore not be neglected.

Keywords Reynolds stress modeling · Spray combustion · Two-way coupling and PDF modeling

---

N.A. Beishuizen (✉)
NUMECA International, Development Engineer, Multiphysics CFD Group, Ch. de la Hulpe 189 Terhulpse Steenweg – 1170 Brussels – Belgium
e-mail: http://www.numeca.be

D. Roekaerts
Department of Multi-Scale Physics, Delft University of Technology, PO Box 5, 2600 AA Delft, The Netherlands
e-mail: d.j.e.m.roekaerts@tudelft.nl

B. Merci et al. (eds.), *Experiments and Numerical Simulations of Diluted Spray Turbulent Combustion*, ERCOFTAC Series 17, DOI 10.1007/978-94-007-1409-0_5,

1 Introduction

In spray flames, the dispersed phase exerts important forces on the carrier medium in two-ways: by the friction between the phases (drag) and by the momentum of the vaporised mass. These forces appear as source terms in the Navier Stokes equations. In turbulent flow, the two-way coupling terms modulate the turbulence characteristics of the gas phase [1, 2]. In turbulent spray flames, the mass transfer can become very important. Fuel vapor coming from the droplets decreases the drag as well as the heat and mass transfer rate of droplets [3].

In this work we use Reynolds stress models and probability density function models to give a statistical description of the above mentioned effects. From the exact multiphase Navier Stokes equations [4–6] the mean momentum, Reynolds stress and turbulence dissipation equation can be derived. In a similar manner, starting again from the Navier Stokes equations, a transport equation for the probability density function (PDF) of velocity (and composition) can be derived [6–8, 48]. The advantage of using the PDF method is that the terms representing convection and chemical reaction appear in closed form and do not have to be modeled.

The solution of the transport equation of the PDF can be obtained using a Lagrangian Monte Carlo method. Lagrangian Monte Carlo simulations for gas phase turbulent reacting flows can be combined with an Eulerian Finite Volume method to achieve more optimal performance [6, 9, 10]. This approach is presented schematically in Fig. 5.1. In this hybrid Monte Carlo/Finite Volume method, developed at TU Delft, mean properties are solved using the Finite Volume method and fluctuating quantities are solved using a Monte Carlo method [49]. The code, called PDFD, calculates the PDF in a 2D axisymmetric domain. Extra care should be taken to assure a consistency between the two components of the hybrid method: the mean Reynolds stresses obtained from the Monte Carlo method should correspond to the mean Reynolds stresses obtained in the Finite Volume method. In particular this concerns the correspondence between the pressure-rate-of-strain term in the Reynolds stress equations and the Langevin model in the Monte Carlo method. The relationship between Lagrangian turbulence models and Eulerian Second Moment Closure models has been investigated in detail for single phase turbulent flows [11, 12, 50].

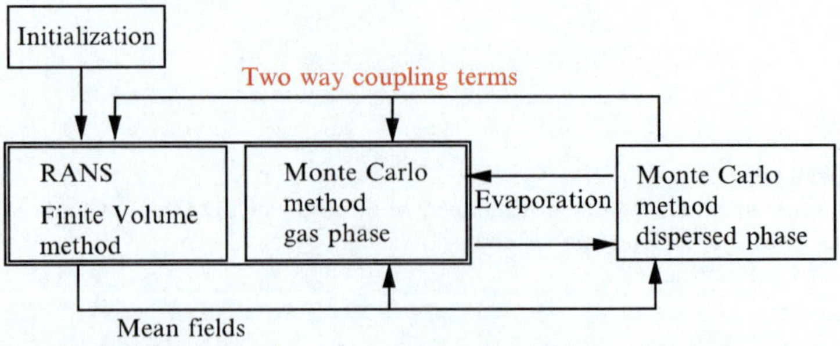

Fig. 5.1 Sketch of the hybrid method

In dispersed two-phase flows, the model for the pressure-rate-of-strain Π_{ij} is modified by the presence of solid particles or droplets. A modified expression for the pressure-rate-of-strain taking into account the presence of the dispersed phase has been derived first by Taulbee et al. [13, 14] following the approach of Launder et al. [15] to obtain an extension of the LRR Reynolds Stress Model. Following the same approach as Speziale et al. [16], a more general formalism was derived by Beishuizen et al. [17]. A straightforward extension can be derived that takes into account mass transfer from droplet vaporization.

In this paper we study the importance of different two-way coupling effects in a turbulent spray flame. The importance of the source term for mass transfer and the importance of the modification of the pressure-rate-of-strain due to the presence of particles is investigated, as well as the modification of the drag and heat transfer coefficient by vaporisation. The developed model is applied to a turbulent non-premixed methanol spray flame studied experimentally by Karpetis and Gomez [18, 19].

2 Theoretical Framework

The mean continuity and mean momentum equations are given by:

$$\frac{\partial \overline{\rho}}{\partial t} + \frac{\partial \overline{\rho}\widetilde{U}_k}{\partial x_k} = \langle S_m \rangle, \tag{5.1}$$

$$\frac{\partial \overline{\rho}\widetilde{U}_i}{\partial t} + \frac{\partial \overline{\rho}\widetilde{U}_k\widetilde{U}_i}{\partial x_k} = -\frac{\partial \overline{\rho}\widetilde{u_i''u_k''}}{\partial x_k} - \frac{\partial \overline{p}}{\partial x_i} + \frac{\partial \overline{\tau_{ik}}}{\partial x_k} + \overline{\rho}g_i + \langle S_{U_i} \rangle + \langle U_{I,i} S_m \rangle, \tag{5.2}$$

where the following notations have been introduced for the mean and the density weighted mean of the transported variable Q:

$$\overline{Q} = \langle Q\mathcal{X} \rangle \quad \text{and} \quad \widetilde{Q} = \frac{\langle \rho Q \mathcal{X} \rangle}{\langle \rho \mathcal{X} \rangle}, \tag{5.3}$$

and the fluctuations

$$q''\mathcal{X} = Q\mathcal{X} - \widetilde{Q}\mathcal{X}. \tag{5.4}$$

The indicator function $\mathcal{X}(\boldsymbol{x}, t)$ of the continuous phase is equal to one when the continuous phase is present at $(\boldsymbol{x}, t)$ and equal to zero otherwise. To be more precise, the notation $\langle\ \rangle$ refers to the expected value of volume averages over small volumes. U_I is the velocity at the interface I between the phases. Considering that the integral of the forces exerted on a droplet surface are equal to the drag force, the mean source term $\langle \mathcal{S}_{U_i} \rangle$ describing momentum transfer at the particle interface is obtained as a sum over the droplets present in a small domain Ω of volume $\mathcal{V}_\Omega$:

$$\langle \mathcal{S}_{U_i} \rangle = -\frac{1}{\mathcal{V}_\Omega} \left\langle \sum_\Omega m_\mathrm{p} \left(\frac{dU_{\mathrm{p},i}}{dt} + \frac{1}{\rho_\mathrm{p}} \frac{\partial \langle p \rangle}{\partial x_i} - g_i \right) \right\rangle, \tag{5.5}$$

where m_p denotes the droplet mass. The mean source term for mass transfer $\langle \mathcal{S}_m \rangle$ is obtained as

$$\langle \mathcal{S}_m \rangle = \frac{1}{\mathcal{V}_\Omega} \left\langle \sum_\Omega \dot{m}_\mathrm{p} \right\rangle. \tag{5.6}$$

The modeled Reynolds stress equation used here is given by:

$$\frac{\partial}{\partial t}\left(\overline{\rho}\widetilde{u_i'' u_j''}\right) + \frac{\partial}{\partial x_k}\left(\overline{\rho}\widetilde{u_i'' u_j''}\widetilde{U}_k\right) = P_{ij} + \mathcal{T}_{ij} + \Pi_{ij} - \frac{2}{3}\overline{\rho}\epsilon\delta_{ij}$$
$$+\langle \mathcal{S}_{u_i u_j} \rangle + 2\langle u_{I,i}'' u_{I,j}'' \mathcal{S}_m \rangle - \widetilde{u_i'' u_j''}\langle \mathcal{S}_m \rangle, \tag{5.7}$$

with

$$\langle \mathcal{S}_{u_i u_j} \rangle = \langle u_{I,i}'' \mathcal{S}_{U_j} \rangle + \langle u_{I,j}'' \mathcal{S}_{U_i} \rangle, \tag{5.8}$$

The contribution to the modulation of the Reynolds stresses due to mass transfer can be decomposed into two terms. First of all, $\left\langle u_{I,i}'' u_{I,j}'' \mathcal{S}_m \right\rangle$ accounts for the effect of mass transfer entering the gas phase with the Reynolds stress $\widetilde{u_{I,i}'' u_{I,j}''}$. The source term $\left\langle u_{I,i}'' u_{I,j}'' \mathcal{S}_m \right\rangle - \widetilde{u_i'' u_j''}\langle \mathcal{S}_m \rangle$ accounts for fluctuations in the (local) vaporisation rate and arises from the continuity equation. This term is zero when the source term for mass transfer shows no local fluctuations: $\mathcal{S}_m' = \mathcal{S}_m - \langle \mathcal{S}_m \rangle = 0$.

The exact production, triple correlations and the pressure-rate-of-strain are given by:

$$P_{ij} = -\overline{\rho}\widetilde{u_i'' u_k''}\frac{\partial \widetilde{U}_j}{\partial x_k} - \overline{\rho}\widetilde{u_j'' u_k''}\frac{\partial \widetilde{U}_i}{\partial x_k}, \tag{5.9}$$

$$\mathcal{T}_{ij} = -\frac{\partial}{\partial x_k}\left[\overline{\rho}\widetilde{u_i'' u_j'' u_k''} - \overline{p' u_i''}\delta_{jk} - \overline{p' u_j''}\delta_{ik}\right], \tag{5.10}$$

$$\Pi_{ij} = \left\langle p' \mathcal{X} \left(\frac{\partial u_i'' \mathcal{X}}{\partial x_j} + \frac{\partial u_j'' \mathcal{X}}{\partial x_i} \right) \right\rangle. \tag{5.11}$$

The production term P_{ij} is in closed form and the triple correlation tensor $\mathcal{T}_{ij}$ is approximated by the Daly-Harlow approximation [20]. The velocity of the gas phase on the interface I is assumed to be equal to the velocity of the particle [21]. The velocity fluctuation at the interface can then be written as:

$$u_{I,i}'' = U_{\mathrm{p},i} - \widetilde{U}_i, \tag{5.12}$$

which was also used by Lain and Sommerfeld [1]. The modeled transport equation for turbulence dissipation is given by the standard equation

$$\frac{\partial \overline{\rho}\epsilon}{\partial t} + \frac{\partial \overline{\rho}\widetilde{U}_j\epsilon}{\partial x_j} = \frac{\partial}{\partial x_j}\left[(\overline{\mu}\delta_{jk} + \frac{C_s^{\epsilon}}{\omega}\overline{\rho}\widetilde{u''_j u''_k})\frac{\partial \epsilon}{\partial x_k}\right]$$
$$+\omega(C_{\epsilon 1}\frac{1}{2}P_{kk} - C_{\epsilon 2}\overline{\rho}\epsilon)$$
$$+\langle S_\epsilon\rangle + \left\langle S_\epsilon^{(m)}\right\rangle. \quad (5.13)$$

The exact two-way coupling source term

$$\langle S_\epsilon\rangle \equiv 2\left\langle \frac{\mu}{\rho}\mathcal{X}\frac{\partial u''_i\mathcal{X}}{\partial x_j}\frac{\partial S_{U_i}}{\partial x_j}\right\rangle. \quad (5.14)$$

is usually modelled as [1,2]

$$\langle S_\epsilon\rangle = C_{\epsilon 3}\frac{\epsilon}{\widetilde{u''_k u''_k}}\langle S_{u_k u_k}\rangle, \quad (5.15)$$

with $C_{\epsilon_3} = 1.8$.

The exact form of the mass transfer source term following from the derivation of the equation for dissipation is given by

$$\left\langle S_\epsilon^{(m)}\right\rangle = 2\left\langle \frac{\mu}{\rho}\mathcal{X}\frac{\partial u''_i\mathcal{X}}{\partial x_j}\frac{\partial u''_i\mathcal{X}}{\partial x_j}S_m\right\rangle. \quad (5.16)$$

When the instantaneous dissipation $\epsilon = 2\frac{\mu}{\rho}\mathcal{X}\frac{\partial u''_i\mathcal{X}}{\partial x_k}\frac{\partial u''_j\mathcal{X}}{\partial x_k}$ and the source term for mass transfer are uncorrelated, this source term can be written in terms of the mean turbulence dissipation *without* a model constant as

$$\left\langle S_\epsilon^{(m)}\right\rangle = \epsilon\langle S_m\rangle. \quad (5.17)$$

2.1 Pressure-Rate-of-Strain Model

Models for the pressure-rate-of-strain tensor Π_{ij} can be constructed by examining the equation for the fluctuating pressure. Following the approach of Speziale et al. [16] to derive the SSG model, the pressure-rate-of-strain takes the form:

$$\Pi_{ij} = \Pi_{ij}^{(1)} + \Pi_{ij}^{(2)}, \quad (5.18)$$

where the slow part of the pressure-rate-of-strain $\Pi_{ij}^{(1)}$ is modelled as:

$$\Pi_{ij}^{(1)} = \overline{\rho} C_1 \epsilon b_{ij} + \overline{\rho} C_2 \epsilon \left(b_{ij}^2 - \frac{1}{3} b_{kk}^2 \delta_{ij} \right), \tag{5.19}$$

and the rapid part $\Pi_{ij}^{(2)}$ as:

$$\begin{aligned} \Pi_{ij}^{(2)} &= \overline{\rho} C_3 k S_{ij} \\ &+ \overline{\rho} C_4 k \left(b_{ik} S_{jk} + b_{jk} S_{ik} - \frac{2}{3} b_{mn} S_{mn} \delta_{ij} \right) \\ &+ \overline{\rho} C_5 k \left(b_{ik} \Omega_{jk} + b_{jk} \Omega_{ik} \right), \end{aligned} \tag{5.20}$$

with $k = \widetilde{u_k'' u_k''}/2$ the turbulent kinetic energy, and with the mean strain rate tensor $S_{ij} = \frac{1}{2}(\frac{\partial \widetilde{U}_i}{\partial x_j} + \frac{\partial \widetilde{U}_j}{\partial x_i})$ and the mean vorticity tensor $\Omega_{ij} = \frac{1}{2}(\frac{\partial \widetilde{U}_i}{\partial x_j} - \frac{\partial \widetilde{U}_j}{\partial x_i})$.

The normalized fluid anisotropy tensor b_{ij} is given by

$$b_{ij} = \frac{\widetilde{u_i'' u_j''}}{\widetilde{u_k'' u_k''}} - \frac{1}{3} \delta_{ij}, \tag{5.21}$$

and the notation b_{ij}^2 is the usual abbreviation for $b_{il} b_{lj}$. The model constants can be chosen in such a way that they correspond to well known turbulence models. Here, we take the constants corresponding to the model of Launder, Reece and Rodi [15].

For dispersed two-phase flows, the model for the pressure-rate-of-strain is augmented by a contribution due to the presence of particles $\Pi_{ij}^{(\mathrm{fp})}$ and a contribution due to vaporisation $\Pi_{ij}^{(m)}$. This extension can be derived using a multiphase version of the approach of Launder, Reece and Rodi [15], or the approach of Speziale, Sarkar and Gatski [16]. More details can be found in the thesis of Beishuizen [17], leading to the model:

$$\Pi_{ij}^{(\mathrm{fp})} = -\frac{1}{2} \langle S_{u_k u_k} \rangle C_1^{(\mathrm{fp})} b_{ij}^{(\mathrm{fp})}, \tag{5.22}$$

with $b_{ij}^{(\mathrm{fp})}$ the normalized anisotropy tensor based on the Reynolds stress two-way coupling source terms:

$$b_{ij}^{(\mathrm{fp})} = \frac{\langle S_{u_i u_j} \rangle}{\langle S_{u_k u_k} \rangle} - \frac{1}{3} \delta_{ij}. \tag{5.23}$$

A similar model contribution appears due to mass transfer, containing the anisotropy tensor $b_{ij}^{(m)}$:

$$\Pi_{ij}^{(m)} = \frac{1}{2} \left\langle u_{I,k}'' u_{I,k}'' S_m \right\rangle C_1^{(m)} b_{ij}^{(m)}, \tag{5.24}$$

$$b_{ij}^{(m)} = \frac{\left\langle u_{I,i}'' u_{I,j}'' S_m \right\rangle}{\left\langle u_{I,k}'' u_{I,k}'' S_m \right\rangle} - \frac{1}{3} \delta_{ij}. \tag{5.25}$$

We assume the model constants $C_1^{(\mathrm{fp})}$ and $C_1^{(m)}$ to be unity, in correspondence with the DNS results of Taulbee et al. [13].

The relative importance of these extra two-way coupling terms appearing in the model for the pressure-rate-of-strain can be seen immediately when they are compared directly with the two-way coupling source terms $\langle S_{u_i u_j} \rangle$ that are explicitly present in the Reynolds stress equations. The source term that is explicitly present in the Reynolds stress equations is given by:

$$\langle S_{u_i u_j} \rangle = \langle u''_{I,i} S_{U_j} \rangle + \langle u''_{I,j} S_{U_i} \rangle, \tag{5.26}$$

and the indirect two-way coupling source term present in the Reynolds stress equations, through the model for Π_{ij}, is given by:

$$\Pi_{ij}^{(\mathrm{fp})} = -\frac{1}{2} C_1^{(\mathrm{fp})} \left(\langle S_{u_i u_j} \rangle - \frac{1}{3} \langle S_{u_k u_k} \rangle \delta_{ij} \right). \tag{5.27}$$

The extra contribution due to the modification of the pressure-rate-of-strain is of the same order of magnitude as the original ('classical') two-way coupling source term.

Using these results, we can now derive a Lagrangian model that corresponds to the model for the pressure-rate-of-strain.

2.2 *Lagrangian Monte Carlo Method*

In the Lagrangian Monte Carlo submodel, the computational gas particle property is described completely by the evolution of the particle location X, the Langevin equation for the particle velocity fluctuation u' and the particle composition $\boldsymbol{\phi}$. The choice of Langevin model implies a model for the pressure-rate-of-strain and these models cannot be chosen independently in the hybrid method. A straightforward extension of the Langevin model that is consistent with the modified model for the pressure-rate-of-strain described earlier is detailed in the PhD thesis of Beishuizen [22]. Here we give the basic outline of the method.

2.3 *Modeling of Fluctuating Acceleration*

Our starting point is the transport equation for the mass density function $\mathcal{F}$ (MDF). To explain the Langevin model it is sufficient to consider a reduced state vector $\boldsymbol{\Psi} = (\boldsymbol{U})$ with $\boldsymbol{U}$ the velocity vector. Substituting the Navier Stokes equations and considering only the continuum phase, we obtain the fluid phase velocity-MDF $\mathcal{F}_U$:

$$\begin{aligned} \frac{\partial \mathcal{F}_U}{\partial t} + V_i \frac{\partial \mathcal{F}_U}{\partial x_i} = &- \frac{\partial}{\partial V_i} [\langle A_i | \boldsymbol{V} \rangle \mathcal{F}_U] \\ &- \frac{\partial}{\partial V_i} \left[\frac{1}{\rho} \langle S_{U_i} | \boldsymbol{V} \rangle \mathcal{F}_U \right] + \frac{1}{\rho} \langle S_m | \boldsymbol{V} \rangle \mathcal{F}_U, \end{aligned} \tag{5.28}$$

where A_i are the forces per unit mass appearing in the Navier Stokes equations:

$$A_i = -\frac{1}{\rho}\frac{\partial p \mathcal{X}}{\partial x_i} + \frac{1}{\rho}\frac{\partial \tau_{ij}\mathcal{X}}{\partial x_j} + g_i \mathcal{X}. \tag{5.29}$$

We first focus on the modelling of the expected value of A_i and then use the same approach to determine the fluid-particle interaction source term and the mass transfer source term. The A_i term is decomposed into a mean and a fluctuating part, using the right-hand side of the Navier-Stokes equation:

$$\frac{\partial}{\partial V_i}[\langle A_i | V\rangle \mathcal{F}_U] = \left(-\frac{1}{\langle \rho\mathcal{X}\rangle}\frac{\partial \langle p\mathcal{X}\rangle}{\partial x_i} + \frac{1}{\langle \rho\mathcal{X}\rangle}\frac{\partial \langle \tau_{ij}\mathcal{X}\rangle}{\partial x_j} + g_i\mathcal{X}\right)\frac{\partial \mathcal{F}_U}{\partial V_i}$$
$$+\frac{\partial}{\partial V_i}\left[\frac{1}{\rho}\left\langle -\frac{\partial p'\mathcal{X}}{\partial x_i} + \frac{\partial \tau'_{ij}\mathcal{X}}{\partial x_j}\middle| V\right\rangle \mathcal{F}_U\right]. \tag{5.30}$$

The expectation of the fluctuating acceleration (conditional upon $\boldsymbol{U}(\boldsymbol{x},t) = \boldsymbol{V}$) needs to be modeled. To do so, we closely follow the approach of Pope [7]. In a small interval of time Δt the position and velocity of the fluid particle at $x(t)$ change by

$$x_i(t+\Delta t) = x_i(t) + U_i(t)\Delta t, \tag{5.31}$$

$$U_i(t+\Delta t) = U_i(t) + \langle A_i\rangle \Delta t + \int_t^{t+\Delta t} a_i(t')dt'. \tag{5.32}$$

By analogy to Langevin's equation, we model the unknown acceleration $\boldsymbol{a}$ by a random contribution and a deterministic contribution. For single phase flows the deterministic part is assumed to be linear in the fluctuating velocity.

For the multiphase interaction terms we assume that the deterministic part is linear in the fluctuation of the particle interaction source term $\mathcal{S}_{U_i}$. For vaporising dispersed phase flows, we also assume that there is a contribution to the deterministic part that is linear in the fluctuation in the source term for mass vaporisation $u''_{I,i}\mathcal{S}_m$. The integral in Eq. 5.32 can then be written as:

$$\int_t^{t+\Delta t} a_i(t')dt' = G_{ij}(U_j - \widetilde{U}_j)\Delta t + B_{ij}dW_j(t) \tag{5.33}$$
$$+G_{ij}^{(\mathrm{fp})}\left(\mathcal{S}_{U_j} - \langle \mathcal{S}_{U_j}\rangle\right)\Delta t$$
$$+G_{ij}^{(m)}\left(u''_{I,j}\mathcal{S}_m - \langle u''_{I,j}\mathcal{S}_m\rangle\right)\Delta t. \tag{5.34}$$

The diffusion term B_{ij} contains three contributions: a fluid phase contribution, a particle interaction contribution and a contribution due to mass transfer:

$$D_{ij} = D_{ij}^{(\mathrm{f})} + D_{ij}^{(\mathrm{fp})} + D_{ij}^{(m)}, \tag{5.35}$$

with $D_{ij} = (B \cdot B^T)_{ij}$. For single phase flows, the stochastic diffusion term $B_{ij}^{(f)}$ is determined based on Kolmogorov's hypothesis [7,23], which states that in isotropic homogeneous turbulence the Lagrangian structure function:

$$D_L^{(f)}(s) \equiv \langle [U^+(t+s) - U^+(t)]^2 \rangle \tag{5.36}$$

for high Reynolds number turbulent flows is of the form:

$$D_L^{(f)}(s) = C_0 \epsilon s, \ \tau_n \ll s \ll T_L. \tag{5.37}$$

In Eq. 5.36, U is a velocity component in any direction, τ_n is the Kolmogorov time scale and T_L is the integral time scale. In Eq. 5.37 C_0 is the Kolmogorov constant and the value of s is in the inertial range. From the model Langevin equation (see e.g. [24]):

$$dU(t) = -U(t)\frac{dt}{T_L} + \left(\frac{2\sigma^2}{T_L}\right)^{\frac{1}{2}} dW(t), \tag{5.38}$$

we obtain the Lagrangian structure function

$$D_L^{(f)}(s) = \frac{2\sigma^2}{T_L} s. \tag{5.39}$$

To be consistent with the Kolmogorov hypothesis, we demand that $\frac{2\sigma^2}{T_l} = C_0\epsilon$. The diffusion term is therefore

$$B_{ij}^{(f)} = \sqrt{C_0 \epsilon} \delta_{ij}. \tag{5.40}$$

The Lagrangian timescale T_L can now be expressed in terms of the turbulent kinetic energy $\widetilde{k} = \frac{1}{2}\widetilde{u_i'' u_i''}$ and the turbulence dissipation ϵ. If we further assume that the velocity variance is that of a Gaussian distribution and therefore given by $\sigma^2 = \frac{2}{3}\widetilde{k}$, the Lagrangian timescale is:

$$T_L^{-1} = \frac{C_0 \epsilon}{2\sigma^2} = \frac{3}{4} C_0 \frac{\epsilon}{\widetilde{k}}. \tag{5.41}$$

2.4 *Modification of the Diffusion Term in Dispersed Two Phase Flows*

We will now propose a model for the diffusion term B_{ij} in Eq. 5.34 to take into account the effect of the presence of the dispersed phase and interphase mass transfer. The derivation starts by a modification of the Lagrangian structure function based on the equation for turbulent kinetic energy valid for dispersed two phase flows. This modification implies a modification of the Kolmogorov hypothesis for the first order Lagrangian structure function. It is known from previous studies that indeed particle laden flows imply a modification of the theoretical results obtained by Kolmogorov, although definitive answers as to how the mechanism of turbulence modification exactly works have yet to be found [25–31].

In multiphase flows, the transport equation for the turbulent kinetic energy can be simplified under the assumption of homogeneous turbulence to yield

$$\frac{\partial \bar{\rho}\widetilde{k}}{\partial t} = -\bar{\rho}\epsilon + \underbrace{\frac{1}{2}\langle S_{u_i u_i}\rangle}_{\langle S_{\mathrm{k}}\rangle} + \underbrace{\langle u''_{I,i} u''_{I,i} S_m\rangle - \frac{1}{2}\langle u''_{I,i} u''_{I,i}\rangle\langle S_m\rangle}_{\langle S_{\mathrm{k}m}\rangle} . \tag{5.42}$$

The right hand side of this equation contains the total dissipation in a turbulent multiphase flow, with a contribution from the fluid phase, the fluid particle interaction and mass transfer. The short hand notations $\langle S_{\mathrm{k}}\rangle$ and $\langle S_{\mathrm{k}m}\rangle$ introduced in Eq. 5.42 will be used from this point to indicate the two-way coupling source terms. Accordingly, the Lagrangian structure function is replaced by:

$$D_L(s) = C_0 \epsilon s - \frac{C_0^{(\mathrm{fp})}}{\bar{\rho}}\langle S_{\mathrm{k}}\rangle s - \frac{C_0^{(m)}}{\bar{\rho}}\langle S_{\mathrm{k}m}\rangle s \tag{5.43}$$

In this equation we have introduced different constants $C_0^{(\mathrm{fp})}$ and $C_0^{(m)}$ for the particle interaction and mass transfer to anticipate on the expected form of the modified turbulence models. The stochastic contribution to the acceleration becomes:

$$\begin{aligned}\int_t^{t+\Delta t} a_i(t')dt' = &\Big[G_{ij}(U_j - \widetilde{U}_j) + G_{ij}^{(\mathrm{fp})}\left(S_{U_j} - \langle S_{U_j}\rangle\right) \\ &+ G_{ij}^{(m)}\left(u''_{I,j} S_m - u''_{I,j}\langle S_m\rangle\right)\Big]\Delta t \\ &+ \left[C_0\epsilon - \frac{C_0^{(\mathrm{fp})}}{\bar{\rho}}\langle S_{\mathrm{k}}\rangle - \frac{C_0^{(m)}}{\bar{\rho}}\langle S_{\mathrm{k}m}\rangle\right]^{\frac{1}{2}} dW_i(t). \end{aligned} \tag{5.44}$$

The last term in Eq. 5.44 is our new model for the stochastic diffusion B_{ij}, including the effects of the presence of particles and the effect of mass transfer.

The corresponding modeled transport equation for the MDF $\mathcal{F}_U$ can be derived [7, 32] and is given by:

$$\begin{aligned}\frac{\partial \mathcal{F}_U}{\partial t} + V_i \frac{\partial \mathcal{F}_U}{\partial x_i} = &-\left(-\frac{1}{\rho(\psi)}\frac{\partial \bar{p}}{\partial x_i} + \frac{1}{\rho(\psi)}\frac{\partial \overline{\tau_{ij}}}{\partial x_j} + g_i\right)\frac{\partial \mathcal{F}_U}{\partial V_i} \\ &- \frac{\partial}{\partial V_i}\left(\mathcal{F}_U\left[G_{il}(V_l - \widetilde{U}_l) - \frac{1}{2}C_0\epsilon\frac{\partial \ln(\mathcal{F}_U)}{\partial V_i}\right]\right) \\ &- \frac{\partial}{\partial V_i}\left(\mathcal{F}_U\left[G_{il}^{(\mathrm{fp})}(S_{V_l} - \langle S_{U_l}\rangle) + \frac{1}{2}\frac{C_0^{(\mathrm{fp})}}{\bar{\rho}}\langle S_{\mathrm{k}}\rangle\frac{\partial \ln(\mathcal{F}_U)}{\partial V_i}\right]\right)\end{aligned}$$

$$-\frac{\partial}{\partial V_i}\left(\mathcal{F}_U\left[G_{il}^{(m)}\left(v''_{I,l}\mathcal{S}_m-\left\langle u''_{I,l}\mathcal{S}_m\right\rangle\right)\right.\right.$$
$$\left.\left.+\frac{1}{2}\frac{C_0^{(m)}}{\overline{\rho}}\langle\mathcal{S}_{km}\rangle\frac{\partial\ln(\mathcal{F}_U)}{\partial V_i}\right]\right). \tag{5.45}$$

It remains to determine a proper model for the drift tensors G_{ij}. When the diffusion and the drift tensors have been determined, it is possible to derive a mean momentum and Reynolds stress equation from them. These equations can then be compared to the Eulerian mean momentum and Reynolds stress equation that were derived by Reynolds averaging the Eulerian Navier Stokes equations.

2.5 Consistency with Second Moment Closures

A modeled transport equation for the Reynolds stresses can be derived from this result by multiplying Eq. 5.45 by $\rho(V_j-\langle U_j\rangle)(V_k-\langle U_k\rangle)$ and integrating. The result is (with a rearrangement of indices to identify the Reynolds stresses $\widetilde{u''_i u''_j}$):

$$\frac{\partial\overline{\rho}\widetilde{u''_i u''_j}}{\partial t}+\frac{\partial\overline{\rho}\widetilde{u''_i u''_j}\widetilde{U}_k}{\partial x_k}+\frac{\partial\widetilde{u''_i u''_j u''_k}}{\partial x_k}+\overline{\rho}\widetilde{u''_j u''_k}\frac{\partial\widetilde{U}_i}{\partial x_k}+\overline{\rho}\widetilde{u''_i u''_k}\frac{\partial\widetilde{U}_j}{\partial x_k}$$
$$=+G_{il}\overline{\rho}\widetilde{u''_j u''_l}+G_{jl}\overline{\rho}\widetilde{u''_i u''_l}+C_0\overline{\rho}\epsilon\delta_{ij}$$
$$+G_{il}^{(\mathrm{fp})}\langle u''_{I,j}\mathcal{S}_{U_l}\rangle+G_{jl}^{(\mathrm{fp})}\langle u''_{I,i}\mathcal{S}_{U_l}\rangle-C_0^{(\mathrm{fp})}\langle\mathcal{S}_{\mathrm{k}}\rangle\delta_{ij}$$
$$+G_{il}^{(m)}\left\langle u''_{I,j}u''_{I,l}\mathcal{S}_m\right\rangle+G_{jl}^{(m)}\left\langle u''_{I,i}u''_{I,l}\mathcal{S}_m\right\rangle-C_0^{(m)}\langle\mathcal{S}_{\mathrm{k}m}\rangle\delta_{ij}.$$

A model for G_{ij}, $G_{ij}^{(\mathrm{fp})}$ and $G_{ij}^{(m)}$ can now be found by comparing the above expression with the Eulerian transport equation for the Reynolds stresses (5.7) [24].

If viscous and pressure transport are neglected, then the model terms correspond to the pressure redistribution term and the dissipation. When dissipation is assumed to be isotropic we obtain the relationship:

$$\Pi_{ij}=G_{il}\overline{\rho}\widetilde{u''_j u''_l}+G_{jl}\overline{\rho}\widetilde{u''_i u''_l}+(1+\frac{3}{2}C_0)\overline{\rho}\epsilon\frac{2}{3}\delta_{ij}$$
$$+G_{il}^{(\mathrm{fp})}\langle u''_{I,j}\mathcal{S}_{U_l}\rangle+G_{jl}^{(\mathrm{fp})}\langle u''_{I,i}\mathcal{S}_{U_l}\rangle-\frac{3}{2}C_0^{(\mathrm{fp})}\langle\mathcal{S}_{\mathrm{k}}\rangle\frac{2}{3}\delta_{ij}$$
$$+G_{il}^{(m)}\left\langle u''_{I,j}u''_{I,l}\mathcal{S}_m\right\rangle+G_{jl}^{(m)}\left\langle u''_{I,i}u''_{I,l}\mathcal{S}_m\right\rangle-\frac{3}{2}C_0^{(m)}\langle\mathcal{S}_{\mathrm{k}m}\rangle\frac{2}{3}\delta_{ij}. \tag{5.46}$$

If we require $\Pi_{ii} = 0$, expressing that the pressure strain term is only redistributive, we obtain the requirement

$$G_{il}\bar{\rho}\widetilde{u_i''u_l''} + \left(1 + \frac{3}{2}C_0\right)\bar{\rho}\epsilon + G_{il}^{(\text{fp})}\langle u_{I,i}'' S_{u_l}\rangle - \frac{3}{2}C_0^{(\text{fp})}\langle S_{\text{k}}\rangle$$
$$+G_{il}^{(m)}\left\langle u_{I,i}'' u_{I,l}'' S_m\right\rangle - \frac{3}{2}C_0^{(m)}\langle S_{\text{k}m}\rangle = 0. \tag{5.47}$$

Because we want this expression to appropriately reduce to the single phase Reynolds stress model as well as the non-vaporising Reynolds stress model, each line in Eq. 5.47 has to be equal to zero independently. We will now first briefly discuss the single phase flow situation and derive an expression for the drift tensor G_{ij}. The particle laden case and the vaporising particle laden case are then subsequently treated in a similar fashion.

2.5.1 Single Phase Flow

The simplest choice for G_{ij} in the case of a single phase flow is

$$G_{ij} = -\frac{1}{2}\left(1 + \frac{3}{2}C_0\right)\frac{\epsilon}{\tilde{k}}\delta_{ij}, \tag{5.48}$$

which is the Simplified Langevin Model (SLM). Substituting this into the first line (the single phase part) of Eq. 5.46 leads to

$$\Pi_{ij}^{(\text{f})} = -\underbrace{\left(1 + \frac{3}{2}C_0\right)}_{c_1}\bar{\rho}\epsilon\left(\underbrace{\frac{\widetilde{u_i''u_j''}}{\tilde{k}} - \frac{2}{3}\delta_{ij}}_{2b_{ij}}\right). \tag{5.49}$$

This corresponds to Rotta's model with the Rotta coefficient $c_1 = 1 + \frac{3}{2}C_0$. A more general expression for G_{ij} can be found by assuming that G_{ij} is a general tensor function depending on n flow properties. Haworth and Pope [33] hypothesized that a sufficiently general functional form for G_{ij} is given by

$$G_{ij} = G_{ij}\left(\widetilde{u_k''u_l''}, \frac{\partial \widetilde{U}_p}{\partial x_q}, \epsilon\right). \tag{5.50}$$

The most general form of this equation that is linear in the anisotropy tensor and in the mean velocity gradients, and that satisfies the four principles of invariant modelling: dimensional consistency, coordinate system independence, Galilean invariance and rules for forming isotropic tensor functions of other tensors, is given by

$$G_{ij} = \frac{\epsilon}{\tilde{k}}(\alpha_1\delta_{ij} + \alpha_2 b_{ij} + \alpha_3 b_{ij}^2) + H_{ijkl}\frac{\partial \widetilde{U}_k}{\partial x_l}. \tag{5.51}$$

2.5.2 Contribution of the Fluid-Particle Interaction

To obtain the most general model for $G_{ij}^{(\mathrm{fp})}$, it is sufficient to consider tensors of the form

$$G_{ij}^{(\mathrm{fp})} = \alpha_1^{(\mathrm{fp})} \delta_{ij} + \alpha_2^{(\mathrm{fp})} b_{ij}^{(\mathrm{fp})} + \alpha_3^{(\mathrm{fp})} \left(b_{ij}^{(\mathrm{fp})} \right)^2 . \tag{5.52}$$

To obtain a model for $G_{ij}^{(\mathrm{fp})}$ that corresponds to the two-way coupling part of the pressure-rate-of-strain tensor $\Pi_{ij}^{(\mathrm{fp})}$ given by Eq. 5.22

$$\Pi_{ij}^{(\mathrm{fp})} = -\frac{1}{2} \langle S_{u_k u_k} \rangle C_1^{(\mathrm{fp})} b_{ij}^{(\mathrm{fp})} \tag{5.53}$$

we can neglect the second and third contribution to the Langevin model in Eq. 5.52. The fluid-particle contribution to the pressure-rate-of-strain reduces to

$$\Pi_{ij}^{(\mathrm{fp})} = \frac{1}{2} C_1^{(\mathrm{fp})} \left(\langle u''_{I,i} S_{U_j} \rangle + \langle u''_{I,j} S_{U_i} \rangle - \frac{2}{3} \left\langle u''_{I,k} S_{u_k} \right\rangle \delta_{ij} \right) . \tag{5.54}$$

A model for the first contribution of the dispersed phase to G_{ij} is then:

$$G_{ij}^{(\mathrm{fp})} = -\frac{1}{2} \cdot \frac{3}{2} C_0^{(\mathrm{fp})} \delta_{ij} . \tag{5.55}$$

It is seen that the contribution due to fluid particle contribution then is similar to a Simplified Langevin Model. Substitution into Eq. 5.46 yields:

$$\Pi_{ij}^{(\mathrm{fp})} = -\frac{3}{2} C_0^{(\mathrm{fp})} \left(\langle u''_{I,i} S_{U_j} \rangle + \langle u''_{I,j} S_{U_i} \rangle - \frac{2}{3} \langle S_{\mathrm{k}} \rangle \delta_{ij} \right) , \tag{5.56}$$

which corresponds to $\Pi_{ij}^{(\mathrm{fp})}$ in the Reynolds stress model (5.53) with $C_1^{(\mathrm{fp})} = 3 C_0^{(\mathrm{fp})}$.

2.5.3 Contribution of Mass Transfer

A model for the mass transfer source term is obtained in a similar manner. The most general model is of the form

$$G_{ij}^{(m)} = \alpha_1^{(m)} \delta_{ij} + \alpha_2^{(m)} b_{ij}^{(m)} + \alpha_3^{(m)} (b_{ij}^{(m)})^2 . \tag{5.57}$$

Choosing $\alpha_1^{(m)} = -\frac{1}{2}\frac{3}{2} C_0^{(m)}$, and $\alpha_2^{(m)} = \alpha_3^{(m)} = 0$, one has

$$G_{ij}^{(m)} = -\frac{1}{2} \cdot \frac{3}{2} C_0^{(m)} \delta_{ij} . \tag{5.58}$$

Substitution into Eq. 5.46 yields:

$$\Pi_{ij}^{(m)} = -\frac{3}{2}C_0^{(m)}\left(2\langle u''_{I,i}u''_{I,j}\mathcal{S}_m\rangle - \frac{2}{3}\langle \mathcal{S}_{\mathrm{k}m}\rangle\delta_{ij}\right). \tag{5.59}$$

There is now a slight inconsistency because the source term $\langle \mathcal{S}_{\mathrm{k}m}\rangle$ contains an extra contribution due to the vaporisation fluctuations which did not appear in the model for the pressure-rate-of-strain derived from the equation for the pressure fluctuation. This inconsistency is most likely caused by the neglect of the mean contributions in the model for the pressure-rate-of-strain, effectively causing the pressure-rate-of-strain to depend on the instantaneous two-way coupling terms instead of the fluctuating two-way coupling terms. Although this inconsistency deserves to be investigated (and solved), for now we will neglect it since it is most likely to play only a small contribution to the total turbulent kinetic energy budget.

2.6 Evolution Equation for Gas Phase Particles

The complete generalised Langevin model for the equation of motion of stochastic samples of the fluid flow in a two-phase turbulent flow is:

$$dX_i(t) = U_i(t)dt, \tag{5.60}$$

$$\begin{aligned} dU_i(t) = &-\frac{1}{\rho}\frac{\partial \overline{p}}{\partial x_i}dt + g_i dt \\ &+G_{ij}(U_j(t) - \widetilde{U_j})dt \\ &+G_{ij}^{(\mathrm{fp})}\left(\mathcal{S}_{U_j} - \langle \mathcal{S}_{U_j}\rangle\right)dt \\ &+G_{ij}^{(m)}\left(u''_{I,j}\mathcal{S}_m - \langle u''_{I,j}\mathcal{S}_m\rangle\right)dt \\ &+\left[C_0\epsilon - \frac{C_0^{(\mathrm{fp})}}{\overline{\rho}}\langle \mathcal{S}_{\mathrm{k}}\rangle - \frac{C_0^{(m)}}{\rho}\langle \mathcal{S}_{\mathrm{k}m}\rangle\right]^{\frac{1}{2}} dW_i(t) \\ &+\frac{1}{\rho}\mathcal{S}_{U_i}dt + \frac{1}{\rho}u''_{I,i}\mathcal{S}_m dt, \end{aligned} \tag{5.61}$$

with G_{ij} given by:

$$\begin{aligned} G_{ij} &= \frac{\epsilon}{\tilde{k}}(\alpha_1\delta_{ij} + \alpha_2 b_{ij} + \alpha_3 b_{ij}^2) + H_{ijkl}\frac{\partial \langle U_k\rangle}{\partial x_l}, \\ G_{ij}^{(\mathrm{fp})} &= -\frac{3}{4}C_0^{(\mathrm{fp})}\delta_{ij}, \\ G_{ij}^{(m)} &= -\frac{3}{4}C_0^{(m)}\delta_{ij}. \end{aligned} \tag{5.62}$$

The models for the fluid particle contribution and mass transfer contribution are Simplified Langevin Models, which correspond to a term similar to a Rotta term (a linear return to isotropy) in second moment closures. It is not necessary to also use a SLM or a Rotta turbulence model for the continuum phase turbulence model. This model can be chosen independently.

In Monte Carlo simulations the mean velocity in a finite volume in the computational domain is obtained by averaging over the N_p Monte Carlo particles present in the cell. The calculated mean velocity will contain a statistical error which is a function of N_p. When the number of particles in the cell is small, the calculated mean velocity will have a large statistical error. When this mean velocity is then used in the calculations of the individual particle evolutions, a *bias* error results, which is a deterministic error that scales with N_p^{-1} [34, 35]. To keep the bias error small, the mean velocities are obtained from the mean transport equations and the fluctuating quantities are obtained by solving a Langevin equation for the fluctuating velocity.

The Langevin equation for the velocity fluctuation $u_i'' = U_i - \widetilde{U}_i$ is:

$$
\begin{aligned}
du_i'' = &- \left(\frac{1}{\rho} - \frac{1}{\overline{\rho}}\right) \frac{\partial \overline{p}}{\partial x_i} dt - u_j'' \frac{\partial \widetilde{U}_i}{\partial x_j} dt + \frac{1}{\overline{\rho}} \frac{\partial \overline{\rho} \widetilde{u_i'' u_j''}}{\partial x_j} dt \\
&+ G_{ij} u_j'' dt \\
&+ (G_{ij}^{(\mathrm{fp})} \left(\mathcal{S}_{U_j} - \langle \mathcal{S}_{U_j} \rangle\right) dt \\
&+ G_{ij}^{(m)} \left(u_{I,j}'' \mathcal{S}_m - \langle u_{I,j}'' \mathcal{S}_m \rangle\right) dt \\
&+ \left(C_0 \epsilon - \frac{C_0^{(\mathrm{fp})}}{\overline{\rho}} \langle \mathcal{S}_\mathrm{k} \rangle - \frac{C_0^{(m)}}{\overline{\rho}} \langle \mathcal{S}_{\mathrm{km}} \rangle \right)^{\frac{1}{2}} dW_i(t) \\
&+ \left[\frac{1}{\rho} \mathcal{S}_{U_i} - \frac{1}{\overline{\rho}} \langle \mathcal{S}_{U_i} \rangle \right] dt + \left[\frac{1}{\rho} u_{I,i}'' \mathcal{S}_m - \frac{1}{\overline{\rho}} \langle u_{I,i}'' \mathcal{S}_m \rangle \right] dt. \qquad (5.63)
\end{aligned}
$$

In Eq. 5.63, the pressure gradient is multiplied by a factor $\frac{1}{\rho} - \frac{1}{\langle \rho \rangle}$. This factor represents the difference between the instantaneous specific volume and the mean specific volume. In an isothermal incompressible flow, the density will be constant and this term vanishes. In a spray flame however, large density variations can occur and this term may not be negligible. For now, this term will be neglected.

In Eq. 5.63 there are also several contributions to the Langevin model due to the presence of dispersed phase particles. The contribution to the mean drift in the third line of Eq. 5.63 and the contribution to the diffusion term in the fifth line of equation result from the modification of the model for the pressure-rate-of-strain. The two-way coupling term in the last line of Eq. 5.63 takes into account the momentum transfer at the interface between the phases, i.e. due to the presence of dispersed phase particles interacting with fluid particles. The mean two-way coupling source terms are known (see Eq. 5.5). But the instantaneous terms $\mathcal{S}_{U_i}$ and $u_{I,i}'' \mathcal{S}_m$ are not known. This follows from the fact that in our simulation method the fluid phase as well as the dispersed phase are both represented by independent samples of the one-point PDF.

Minier and Peirano [36] have already proposed a stochastic closure model for these instantaneous coupling terms between the phases taking into account the dispersed phase volume fraction. When the Langevin equation is of the form

$$dU_i = A_{\mathrm{f},i}\,dt + A_{\mathrm{p}\to\mathrm{f},i}\,dt + B_{ij}\,dW_j, \tag{5.64}$$

in their approach the two-way coupling source term $A_{\mathrm{p}\to\mathrm{f}}$ is given by

$$A_{\mathrm{p}\to\mathrm{f},i} = \begin{cases} 0 & \text{with probability } 1-\alpha_\mathrm{p} \\ \frac{1}{\rho_\mathrm{f}} S_{U_i} = \frac{1}{\rho_\mathrm{f}\mathcal{V}_\mathrm{f}} F_{\mathrm{p}\to\mathrm{f},i} & \text{with probability } \alpha_\mathrm{p} \end{cases} \tag{5.65}$$

where $\mathcal{V}_\mathrm{f}$ is the volume occupied by fluid particle f. The force exerted by a dispersed phase particle on the surrounding fluid $F_{\mathrm{p}\to\mathrm{f}}$ is the right hand side of the equation of motion of the particle, $F = m\frac{dU_\mathrm{p}}{dt}$, which is given in Eq. 5.67. When such force is needed in the equation of motion of a gas phase particle, according (5.65), it is drawn from the distribution formed from the dispersed particles in the neighborhood.

More recently Anand and Jenny have developed a unified joint PDF framework [10, 51] for dilute dispersed multiphase flow. In their approach the joint PDF of continuous and dispersed phase properties is represented by a large ensemble of computational particles, with each particle carrying information of both phases. The terms in the gas phase particle equation referring to instantaneous droplet properties then are explicitly represented.

We here use a simpler model. It is an extension of the algorithm used for mass transfer. The distribution algorithms for the source terms of mass transfer can be constructed using the saturation of the fluid phase particles [6] as a parameter to determine which gas phase particles receive mass from dispersed phase particles. The idea is based on the observation that evaporation should feed a peak at saturated conditions in the gas phase composition PDF [37]. We sort the fluid particles in a finite volume cell based on their saturation conditions. The fluid particle that is closest to saturation gets vaporised mass until it reaches saturation conditions. We then move to the next particle that is closest to saturation until all the vaporised mass is given to the fluid particles. This algorithm is described in more detail in [6]. In a straightforward extension, instantaneous momentum transfer is assumed to happen between the gas phase particles and the disperse phase particles that are already associated with each other in the frame of the mass transfer algorithm. The evolution of the fluid particle composition $\boldsymbol{\phi}$ is governed by the addition of evaporated liquid fuel mass by the algorithm described above, by chemical reactions and by a Lagrangian micromixing model. Here, the standard Interaction by Exchange with the Mean (IEM) mixing model is used [38].

In the example studied below, only a single independent variable Z, the mixture fraction, is considered to describe the evolution of the gas phase composition $\boldsymbol{\phi}$: $\boldsymbol{\phi} = (Z)$. A flamelet lookup table is used to obtain the thermochemical properties of the gas phase as a function of the mixture fraction Z. In the evolution equation for particle mixture fraction there is no chemical source term. The strain rate is known to vary throughout the domain in reacting flows. However, it is fixed at a value of $s = 100$ 1/s in these simulations. Simulations of a spray flame where variable strain rate is taken into account (through the scalar dissipation rate) can be found in the works of Ge and Gutheil, i.e. [39]. In the transport equation for the scalar dissipation rate, the two-way coupling source terms also appear and therefore the mixing model is in principle influenced by the presence of particles and vaporisation. A simple analysis following Fox [38] shows that the effect of mass transfer on the scalar dissipation rate is a function of $Re^{-\frac{1}{2}}$ and is therefore negligible at high Reynolds numbers. If the model constant for the two-way coupling source term is not a function of the Reynolds number, then this term is also negligible for high Reynolds numbers and the mixing model frequency is unaltered by the vaporising droplets.

2.7 *Governing Equations for the Dispersed Phase*

The dispersed phase consists of liquid fuel droplets that move, heat up and vaporise as they traverse downstream. They are considered point particles as the flow in and around each droplet is not resolved, but the volume occupied by the droplets as well as the forces exerted by the particles on the flow is taken into account through the volume fraction and the two-way coupling terms in the continuum phase equations. Since the focus of the article is on two-way coupling, a full description of the governing equations for the dispersed phase is given below.

2.7.1 Acceleration Model

Simplified equations of motion are used to describe the velocity evolution of a heavy particle with mass m_p in a turbulent fluid [40]:

$$\frac{\mathrm{d}X_{\mathrm{p},i}}{\mathrm{d}t} = U_{\mathrm{p},i}\,, \tag{5.66}$$

$$\frac{\mathrm{d}U_{\mathrm{p},i}}{\mathrm{d}t} = \frac{U_{\mathrm{s},i} - U_{\mathrm{p},i}}{\tau_\mathrm{p}} - \frac{1}{\rho_\mathrm{p}}\frac{\partial \langle p \rangle}{\partial x_i} + g_i\,. \tag{5.67}$$

The effect of the surrounding fluid flow is included through the first two terms on the right hand side of Eq. 5.67, respectively the drag force and the mean pressure gradient at the particle location. The seen velocity $\boldsymbol{U}_\mathrm{s}$ is the velocity of the undisturbed fluid flow at the position of the particle center – the velocity that would exist in the absence of the particle but turbulent and disturbed by all the other particles [41]. It is modeled using the stochastic dispersion model of Minier and Peirano [6,22,36] which takes into account the effect of crossing trajectories and the continuity effect. The final Langevin equation for the velocity seen, including two-way coupling, is:

$$
\begin{aligned}
dU_{s,i} = & -\frac{1}{\rho_f}\frac{\partial \overline{p}}{\partial x_i}dt + \langle U_{r,i}\rangle \frac{\partial \langle U_i \rangle}{\partial x_j}dt \\
& - \left(\frac{1}{2}+\frac{3}{4}C_0\right)\frac{\epsilon}{k}b_i\left(U_{s,i}-\langle U_{s,i}\rangle\right)dt + \sqrt{\frac{4}{3}\epsilon\left(\left(\frac{1}{2}+\frac{3}{4}C_0\right)b_i\frac{\mathcal{K}}{k}-\frac{1}{2}\right)}dW_i \\
& + \frac{1}{\rho_f}S_{U_i}dt + \frac{1}{\rho_f}U_{f,i}S_m dt,
\end{aligned} \tag{5.68}
$$

In the case of anisotropic turbulence a weighted turbulent kinetic energy has to be introduced to compensate for the non-isotropic drift vector:

$$
\mathcal{K} = \frac{3}{2}\frac{\sum_{i=1}^{3} b_i \left\langle u'_{f,i}u'_{f,i}\right\rangle}{\sum_{i=1}^{3} b_i}, \tag{5.69}
$$

with the ratio of the fluid timescale and the seen timescale $b_i = \frac{T_f}{T_i}$ [36].

In Eq. 5.67 the particle response time scale τ_p is based on the response time in a Stokes regime $\tau_p^{(St)}$:

$$
\frac{1}{\tau_p} = \frac{\mathrm{Re}_p}{24}\frac{C_D}{\tau_p^{(St)}} \quad \text{with} \quad \tau_p^{(St)} = \frac{\rho_p d_p^2}{18\mu_f}, \tag{5.70}
$$

where C_D is the Schiller-Naumann drag coefficient for high Reynolds number flows:

$$
C_D = \begin{cases} \frac{24}{\mathrm{Re}_p}\left(1+0.15\mathrm{Re}_p^{0.687}\right) & \text{if } \mathrm{Re}_p \le 800 \\ 0.44 & \text{if } \mathrm{Re}_p > 800 \end{cases} \tag{5.71}
$$

with $\mathrm{Re}_p = \rho_f \left|\boldsymbol{U}_p - \boldsymbol{U}_s\right| d_p \mu_f$.

2.7.2 Evaporation Model

The evolution of droplet mass and droplet temperature is described by an infinite conductivity model. We use here the non-equilibrium model proposed by Bellan and Harstad [42] and the formalism described in Miller et al. [43]. The evolution of droplet mass and temperature is then given by:

$$
\dot{m}_p = -\frac{1}{3}\frac{\mathrm{Sh}}{\mathrm{Sc}}\frac{m_p}{\tau_p}\ln(1+B_M), \tag{5.72}
$$

$$
\dot{T}_p = \frac{f_2\mathrm{Nu}}{3\mathrm{Pr}}\frac{C_{pf}}{C_{pl}}\frac{(T_f - T_p)}{\tau_p^{St}} - \frac{\dot{m}_p}{m_p}\frac{L}{C_{pl}}, \tag{5.73}
$$

with L the latent heat of vaporisation. The Nusselt number Nu and the Sherwood number Sh are given by the Ranz-Marshall correlations:

$$\mathrm{Nu} = 2 + 0.552\mathrm{Re}_\mathrm{p}^{1/2}\mathrm{Pr}^{1/3}, \tag{5.74}$$

$$\mathrm{Sh} = 2 + 0.552\mathrm{Re}_\mathrm{p}^{1/2}\mathrm{Sc}^{1/3}, \tag{5.75}$$

where the Prandtl number $\mathrm{Pr} = \mu_\mathrm{f} c_{p,\mathrm{f}} / \lambda_\mathrm{f}$ and the Schmidt number $\mathrm{Sc} = \mu_\mathrm{f} / (\rho_\mathrm{f} \mathbb{D}_\mathrm{f})$. The Spalding mass transfer number and the Spalding heat transfer number are defined as

$$B_M = \frac{Y^{(I)} - Y_\mathrm{f}}{1 - Y^{(I)}}. \tag{5.76}$$

In Eq. 5.76, $Y^{(I)}$ denotes the vapor mass fraction at the droplet interface and Y_f is the local vapor mass fraction of the surrounding gas phase (the mean mass fraction in the finite volume cell in which the particle resides). We have used the classical rapid mixing model in combination with a Clausius Clapeyron relationship providing the equilibrium vapor mole fraction and with the non-equilibrium model of Miller et al. [43]. No significant differences were found in the simulation results when comparing these two models and results presented below are for the model of Miller unless stated otherwise.

2.7.3 Drag and Heat Transfer Modification Due to Vaporisation

Vaporizing droplets have a smaller drag coefficient than non-vaporizing droplets. The drag coefficient is modified by the Spalding mass transfer number B_M or Spalding heat transfer number B_T. We use here the drag correction proposed by Chiang et al. [44]:

$$C_{D_\mathrm{evap}} = C_{D_\mathrm{sphere}} (1 + B_H)^{-0.32}, \tag{5.77}$$

with

$$B_H = B_T \left(1 - \frac{Q_l}{Q_\mathrm{f}}\right). \tag{5.78}$$

In Eq. 5.78, Q_l is the energy consumption rate for droplet heating and Q_f is the total heat flux obtained from the gas phase. B_T is the Spalding heat transfer number

$$B_T = \frac{C_{p_\mathrm{f}}(T_{\mathrm{f}\infty} - T_\mathrm{f}^{(I)})}{L_\mathrm{eff}} \tag{5.79}$$

In Eq. 5.79, $T_{indf}^{(I)}$ denotes the temperature at the droplet interface and $T_{\mathrm{f}\infty}$ is the local temperature of the surrounding gas phase. Chiang also finds correlations for the Nusselt and Sherwood number for vaporizing droplets using the heat transfer number B_H and B_M:

$$\mathrm{Nu} = \mathrm{Nu}_\mathrm{Sphere} (1 + B_H)^{-0.700}, \tag{5.80}$$

$$\mathrm{Sh} = \mathrm{Sh}_\mathrm{Sphere} (1 + B_M)^{-0.557}, \tag{5.81}$$

leading to lower Nusselt and Sherwood numbers compared to the classical Ranz-Marshall results. These corrections incorporate the slower droplet vaporisation due to the finite diffusion of the fuel vapor to the environment.

3 Spray Flame Simulations

A turbulent methanol spray flame under conditions similar to those measured experimentally in the experiments of Karpetis and Gomez [18,19,45] has been numerically simulated. This experiment will be referred to as the Karpetis spray. A sketch of the experimental configuration with the numerical domain can be found in Fig. 5.2. The spray flame is of the non-premixed type with a liquid mass flow rate of 5.1 g/min. The Reynolds number, based on cold flow conditions, was approximately 21,000 and the Kolmogorov length scale was estimated to be $\eta = 3\ \mu$m.

Figure 5.2 shows the computational domain used and the location of the measurement sections. Measurements were performed at $x = 0.1D$ (used as inlet conditions in the simulations), $x = 0.5D$ and at every half burner diameter D ($D = 12.7$ mm) thereafter, up to $x = 6.5D$. Phase Doppler Anemometry (PDA) was used to obtain the first and second moments of gas and droplet velocities, droplet number density and droplet size. Results for three different size classes, together with the mean

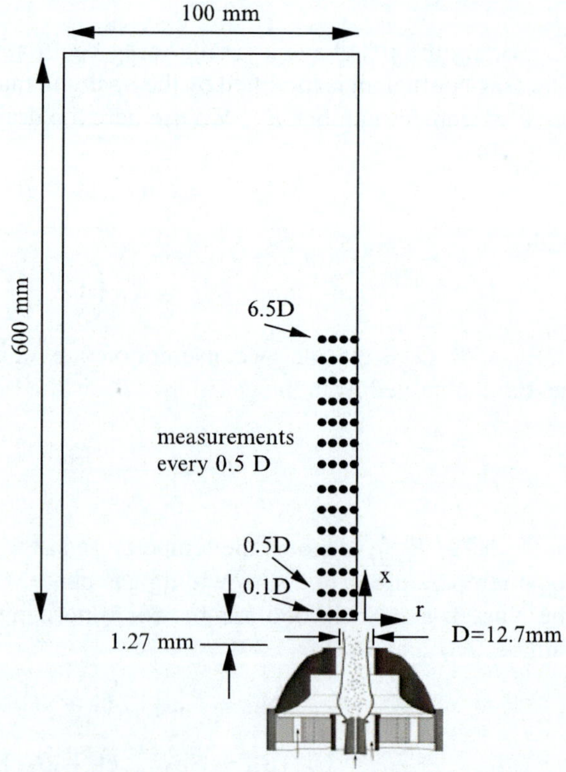

Fig. 5.2 Sketch of the configuration used by Karpetis and Gomez, and the numerical domain used in the simulations presented here

results, are available. The gas velocity statistics were calculated by considering all droplets smaller than 8 μm as tracer particles of the gas flow. This information is also used in the dispersed phase as a fourth size class. The Stokes number is a measure for the response time of a droplet and is defined as: $\mathrm{St} = \tau_\mathrm{p}/\tau_\mathrm{f} = \rho_\mathrm{p} d_\mathrm{p}^2/18\nu_\mathrm{f}\rho_\mathrm{f}\tau_\mathrm{f}$. The Stokes number based on the flow timescale $\tau_\mathrm{f} = D/U$ is $\mathrm{St} = 0.05$ for the smallest droplets (the Stokes number based on the Kolmogorov timescale is $\mathrm{St} = 0.66$). It is mentioned in [45] that tests where either 6 or 12 μm was used as an upper limit yielded minimal differences in the velocity and velocity correlations attributed to the gas phase, which indicate that the droplet size class is small enough to be considered as tracer particles.

In addition to the PDA measurements, Raman Spectroscopy was used to measure mean gas temperature and temperature fluctuations. At $x = 0.1D$, the spray is already burning, which means that part of the droplets have already evaporated to form a reacting mixture. The mixture fraction profile at $x = 0.1D$ is chosen in such a way that the corresponding flamelet temperature matches the measured mean temperature profile. The flamelet temperature corresponding to the most lean condition was used. The pressure is solved using a pressure correction algorithm, which takes into account mass transfer.

3.1 Dispersed Phase

For the dispersed phase, experimental data on several droplet size classes are available. Nine droplet size classes were used in the simulations. The first two size classes are [0–8] and [8–15] μm, the other classes each have a bandwidth of 10 μm, starting from [15–25] μm and ending with [75–85] μm. The mean droplet diameter was $d_\mathrm{p} = 30\ \mu$m, which is almost the same size as the Kolmogorov length scale. Direct numerical simulations of Elghobashi and Druzhinin have shown that the turbulence dissipation increases when the droplet diameter d_p is larger than the Kolmogorov length scale η_K and decreases when $d_\mathrm{p} < \eta_K$. This behaviour is caused by interaction between different length scales in the energy cascade process and can therefore not be observed in Reynolds stress models where only a single length scale of turbulence is present. It is for this reason that it is difficult to correctly predict the turbulence dissipation in polydispersed phase flows with two-way coupling.

The liquid droplets enter the domain with a temperature of $T_\mathrm{p} = 300$ K for all droplets. This temperature then quickly increases towards the boiling temperature $T = 334$ K.

Thirty computational gas-phase particles per cell were used and six computational liquid-phase particles per class per cell were used (which means a total of 54 computational liquid-phase particles per cell). A comparison of a simulation where 60 computational gas-phase particles and 9 liquid-phase particles were used with a simulation where 30 gas particles and 3 liquid particles were used has shown a difference of less than 2% in the velocity of the smallest particles and less than 10% in the Reynolds stresses. The difference in Reynolds stresses of the gas-particles was approximately 5% and less than 2% near the centerline. Of more influence was the time step used for the dispersed phase droplets. A local timestepping method [46] was used for the gas phase and the liquid phase with timesteps varying around $\Delta t = 1.0 \cdot 10^{-6}$ s. Choosing a smaller time step had a negligible influence on the gas phase results, but it takes much more iterations to obtain smooth dispersed phase averages.

The coupled equations of motion (5.66), (5.67) and (5.68) are solved using a weak first order stochastic scheme based on the stochastic scheme of Minier, Peirano and Chibbaro [47], but extended for the case of a generalised coordinate system where the mean drift vector need not be aligned with the first axis of the coordinate system [22]. The main result is that stochastic terms also enter the propagation equation for $X_{\mathrm{p},i}$ and $U_{\mathrm{p},i}$.

The total number of timesteps used was 30,000, which took approximately 2 days on a 3 GHz 64 bit single processor.

4 Results and Discussion

4.1 Gas Phase Results

Figure 5.3 shows radial profiles of the mean axial velocity of the gas phase $\widetilde{U}_{\mathrm{f}}$ at four axial locations downstream of the inlet. The dots represent the experimental data of the smallest droplet size class (0–8 μm), which is used as a representation of the gas phase velocity.

The momentum transfer of the droplets will reduce $\widetilde{U}_{\mathrm{f}}$ because $\widetilde{U}_{\mathrm{p}} < \widetilde{U}_{\mathrm{f}}$, but the mass transfer due to vaporisation will increase $\widetilde{U}_{\mathrm{f}}$. It can be seen in Fig. 5.3 that the latter effect is slightly stronger resulting in higher axial velocities near the centerline. In these simulations is important to take into account that vaporisation reduces the drag coefficient and the heat and mass transfer coefficient with a factor of $1 + B_M$. This effect was taken into account in the vaporisation correction of Chiang [44] (see Sect. 2.7.3). The momentum transfer and therefore the mean axial velocity are lower when this effect is not included. From this point on, all simulation results include the effect of vaporisation on the drag and heat and mass transfer coefficient.

We will now look at four different simulations, referred in the text as cases (a) through (d), with different settings for the two-way coupling source terms (Table 5.1). Case (a) does not include any two-way coupling effects. Case (b) includes all two-way coupling effects in the momentum and Reynolds stress and dissipation equation, except those in the model for the pressure-rate-of-strain. Case (c) is like case (b), but it also includes the effect of all mass transfer terms in the momentum, Reynolds stress and dissipation equation, and case (d) is like case (c) but including the effect of particles and mass transfer on the pressure-rate-of-strain Π_{ij}. The mean axial velocity is shown in Fig. 5.4. Including two-way coupling always leads to a lower mean axial velocity in the center of the spray, and all simulation results are lower than the experimental data here.

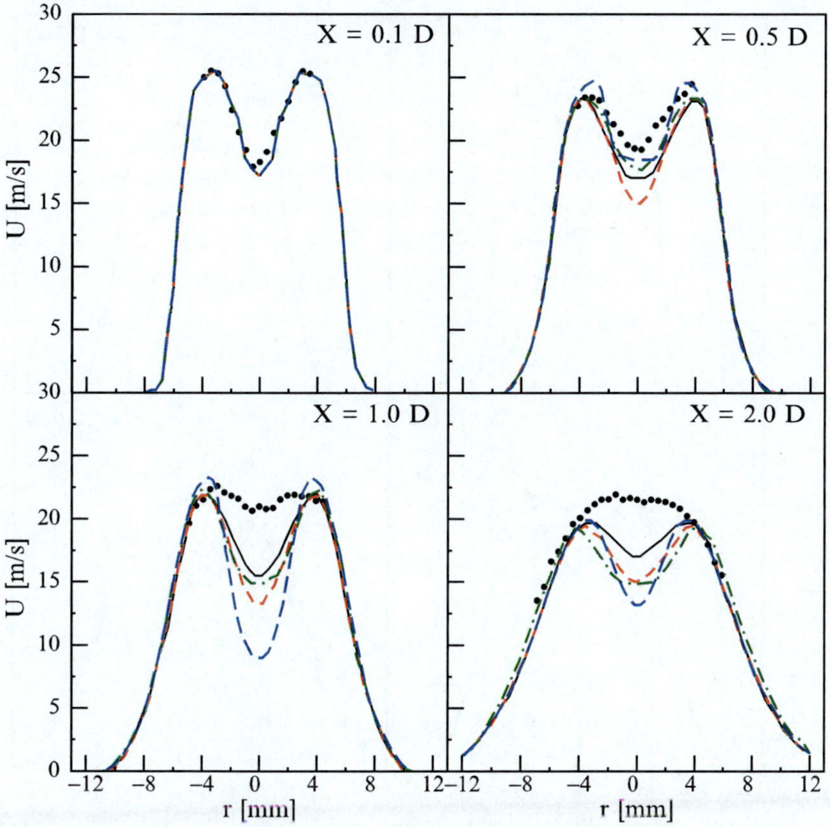

Fig. 5.3 Comparison of two-way coupling models with the correction model of Chiang. • Exp. data Karpetis. —— No two-way coupling. - - - one-way coupling. -·- one-way + mass coupling. — — Without Chiang correction

Table 5.1 The source terms that are taken into account in the different simulations

Case	Source terms
(a)	[-]
(b)	$\mathcal{S}_{U_i}, \mathcal{S}_{u_iu_j}, \mathcal{S}_\epsilon$
(c)	$\mathcal{S}_{U_i}, \mathcal{S}_{u_iu_j}, \mathcal{S}_\epsilon$
	$\mathcal{S}_m, U_{\mathrm{p},i}\mathcal{S}_m, u_i''u_j''\mathcal{S}_m, \mathcal{S}_{\epsilon m}$
(d)	$\mathcal{S}_{U_i}, \mathcal{S}_{u_iu_j}, \mathcal{S}_\epsilon, \Pi_{ij}^{(\mathrm{fp})}$
	$\mathcal{S}_m, U_{\mathrm{p},i}\mathcal{S}_m, u_i''u_j''\mathcal{S}_m, \mathcal{S}_{\epsilon m}, \Pi_{ij}^{(m)}$

The axial Reynolds stress $\widetilde{u_\mathrm{f}''u_\mathrm{f}''}$ is shown in Fig. 5.5. A depression near the centerline can be observed in the experiment and the simulation without two-way coupling (case (a)). Two way coupling leads to an increase in the Reynolds stress. Including two-way coupling in the model for Π_{ij} (case (d)) leads to a partial compensation of the total two-way coupling source terms and results in a lower Reynolds stress with respect to cases (b,c). Also, the contribution of mass transfer is not negligible and should be taken into account in the Reynolds stress equations. The radial Reynolds stress $\widetilde{v_\mathrm{f}''v_\mathrm{f}''}$ is shown in Fig. 5.6. When no two-way coupling is used, the

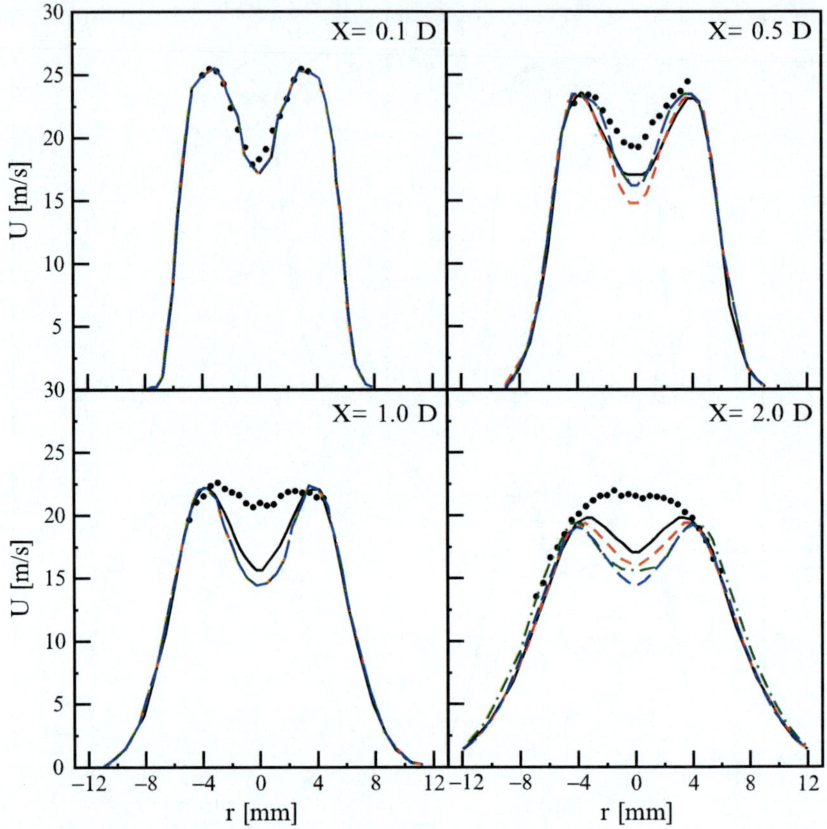

Fig. 5.4 Radial profiles of mean gas phase axial velocity $\widetilde{U}_f$. • Exp. data Karpetis. —— No two-way coupling. - - - Classic two-way coupling. -·- two-way + mass coupling. — — Full two-way coupling

experimental profile and the simulated profile for the Reynolds stress is qualitatively and quantitatively the same. Including two-way coupling effects leads to a different Reynolds stress profile. Including two-way coupling slightly increases the radial Reynolds stress, causing a local maximum at the centerline. Including mass transfer has a much stronger effect, causing a double peak profile around the centerline. The mass transfer source term is slightly less strong at the centerline because the temperature and therefore vaporisation is slightly lower there, which explains the shape of the profile. Again, including the effect of the presence of droplets and mass transfer on the pressure-rate-of-strain slightly reduces the strength of the two-way coupling source term, leading to a lower peak compared to case (c). At a distance of $X = 2.0D$ downstream of the injector, the 'full' two-way coupling model has a Reynolds stress profile that is flatter than the other cases.

Figure 5.7 shows a comparison between the experimentally measured mean temperature profile and the simulation results. All simulations show a local minimum at the centerline: the temperature at the centerline drops due to excess fuel vapor, and this a sign of (at least numerically) of external group combustion. The experiment

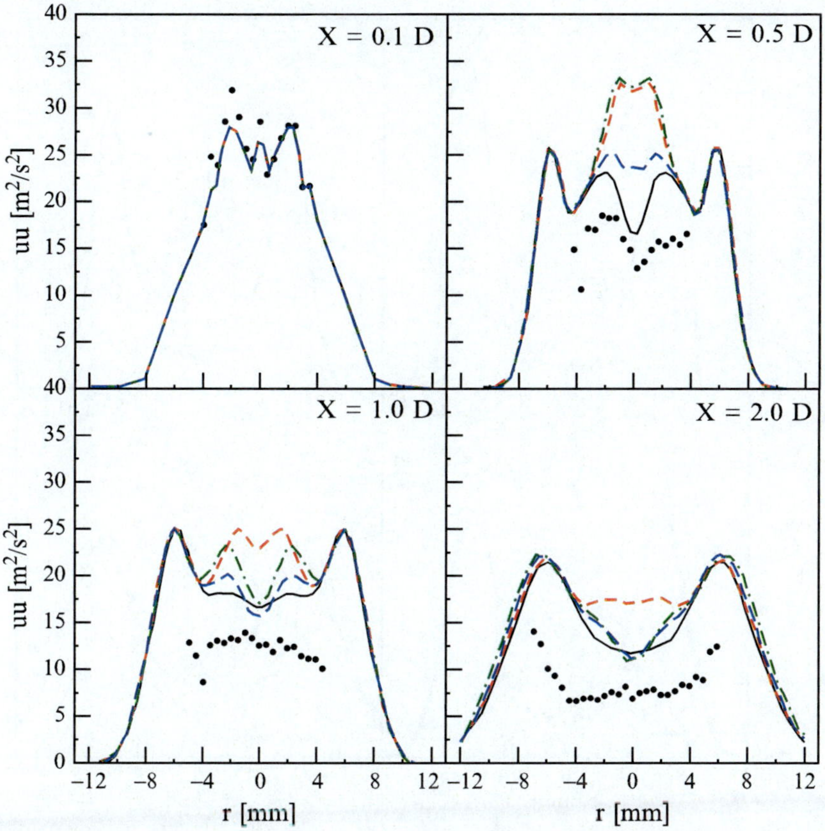

Fig. 5.5 Gas phase axial Reynolds stress $\widetilde{u_f'' u_f''}$. • Exp. data Karpetis. — No two-way coupling. --- Classic two-way coupling. -.- 2-way + mass coupling. — Full 2-way coupling

does not show this behavior. This could be an experimental error [18]. The temperature was only measured when the measurement was not disturbed by the presence of a particle. However, the presence of a vaporizing particle will mean a local increase in fuel mass fraction and therefore a lower temperature. Uncertainties in the modeling can also be the cause of the discrepancy. It is therefore also interesting to look at the temperature variance, shown in Fig. 5.8. All simulation results show a very distinct minimum at the centerline, even the simulation without two-way coupling. The fact that the fluctuations in temperature are damped can be related to the algorithm for mass transfer between Monte Carlo particles. When the fuel vapour emanating from liquid droplets enters the gas phase, it is at (approximately) the boiling temperature of methanol. This fuel vapour is then given to the Monte Carlo particles in a 'most saturated particle first' approach [6]. Immediately afterwards, the flamelet table assigns a new temperature to the Monte Carlo particle. When all particles are saturated, the remaining vapour is distributed homogeneously over all gas-particles in the finite volume cell, leading to small or no variance.

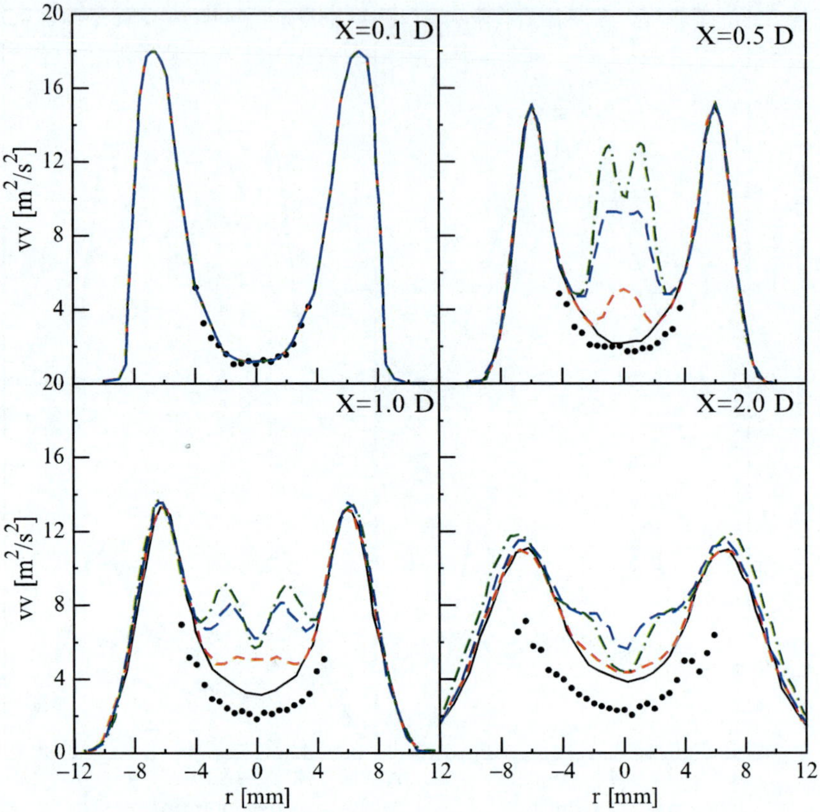

Fig. 5.6 Gas phase axial Reynolds stress $\widetilde{v_f'' v_f''}$. • Exp. data Karpetis. — No two-way coupling. - - - Classic two-way coupling. -·- 2-way + mass coupling. — — Full 2-way coupling

4.2 Dispersed Phase Results

Figure 5.9 shows the axial velocity of the dispersed phase for three different droplet size classes. The smallest droplet size class was used to represent the gas phase velocity in the previous figures. The simulation results are for case (d) only, including all two-way coupling effects. As reference, the simulation results for the gas phase are also included. The difference in velocity between the different size classes is captured very well in the simulations, although further downstream the velocity profiles of the smaller droplet sizes do not develop as fast as in the experiment. The difference between the velocity of the smallest droplet size class and the gas phase velocity causes a problem: the smallest droplets are, at least in the simulations, not a good representation of the gas phase velocity. This is true even when no two-way coupling is used in the simulations, and should be attributed to the particle vaporization. The smallest droplets have only been small for a short amount of time and they did not have time yet to relax to the gas phase velocity. They will be evaporated before their velocity has fully relaxed to the gas phase.

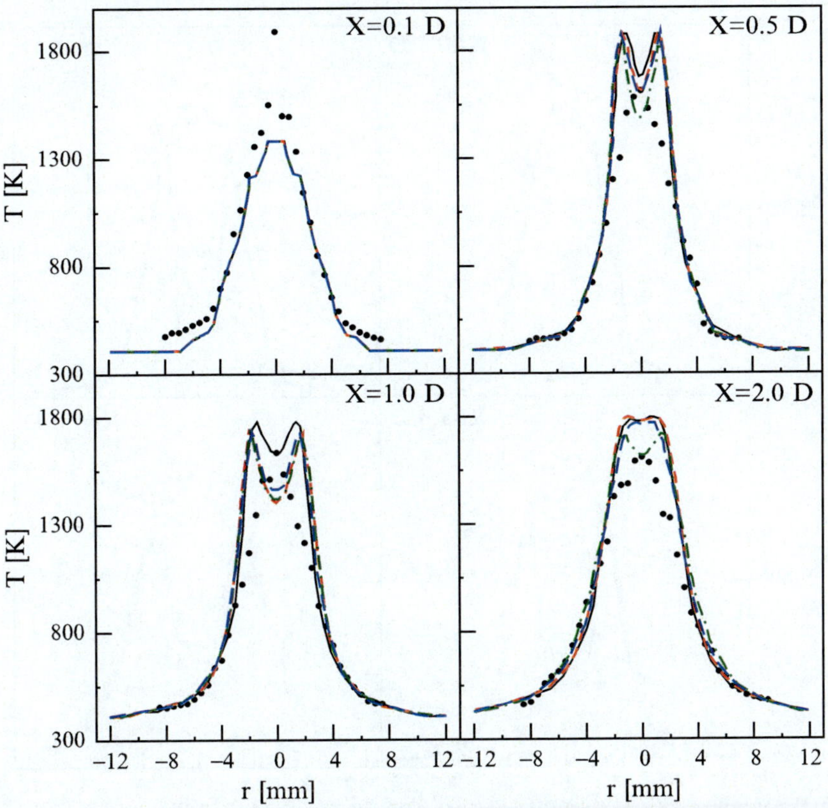

Fig. 5.7 Gas phase temperature $\widetilde{T}_{\mathrm{f}}$. • Exp. data Karpetis. — No two-way coupling. - - - Classic two-way coupling. –·– 2-way + mass coupling. — — Full 2-way coupling

A comparison with the radial velocity shows that the droplet velocity is also different from the gas phase velocity, but for a different reason. Figure 5.10 shows a comparison between the measured radial velocity of the gas phase, the simulated radial velocity of the gas phase and the simulated radial velocity of the droplets in the size class [0–8] μm. The figure clearly shows that the computed radial velocity of the smallest droplets agrees very well with the experimentally observed values of radial velocity. Note that the experimentally measured values are in fact the measured radial velocities of the smallest droplets.

An explanation for this is as follows: a tracer liquid particle originally situated in the spray region will have a large positive radial velocity and it will therefore move away from the centreline. A fluid particle situated outside of the spray region has a much lower radial velocity. However, the velocity of this fluid particle cannot be measured since the flow surrounding the spray region does not contain any particles. The velocity measurements near the edge of the spray are biased due to the conditional measuring of tracer particles coming from the spray region.

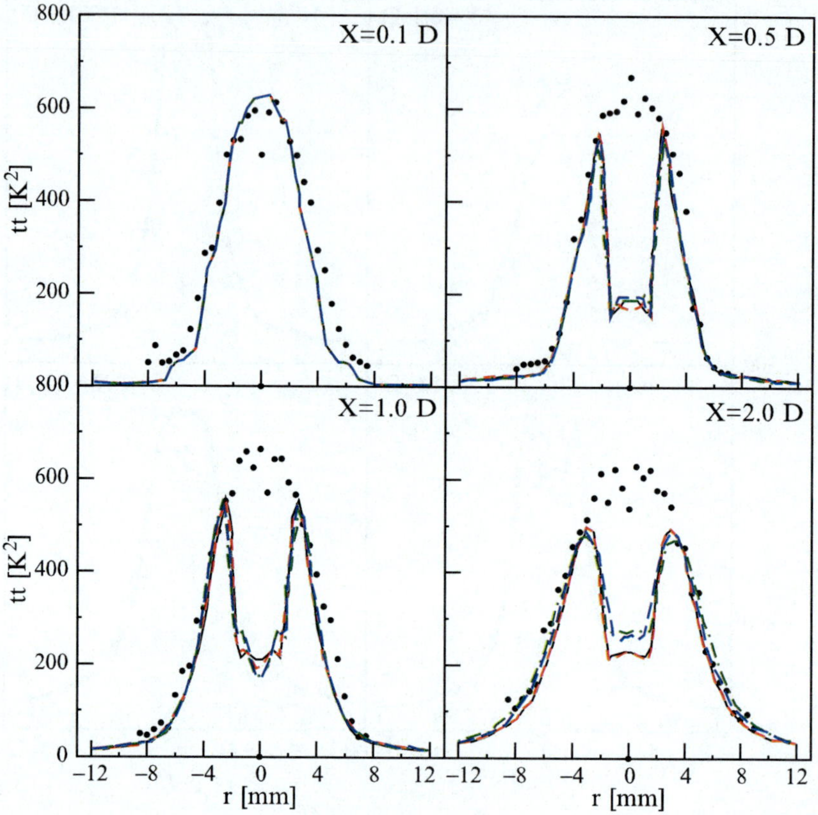

Fig. 5.8 Temperature variance $\sqrt{\widetilde{T_f''T_f''}}$. • Exp. data Karpetis. — No two-way coupling. - - - Classic two-way coupling. –·– 2-way + mass coupling. — — Full 2-way coupling

In Fig. 5.11 the droplet radial Reynolds stress is shown, together with the simulation results for the gas phase. The Reynolds stress is predicted reasonably well, although at the edge of the spray region there is a tendency to overpredict the Reynolds stress for the smaller droplet size classes.

Figure 5.12 shows the mean diameter. The results without two-way coupling and with full two-way coupling are shown here. The differences for the predictions of the mean droplet diameter are marginal because the temperature and axial velocity are almost the same for both simulations.

5 Conclusions

In a spray flame many interaction effects can play a role. This work has focused on two-way coupling effects due to momentum and mass transfer. These effects were studied in the framework of a hybrid method, combining a Reynolds stress model

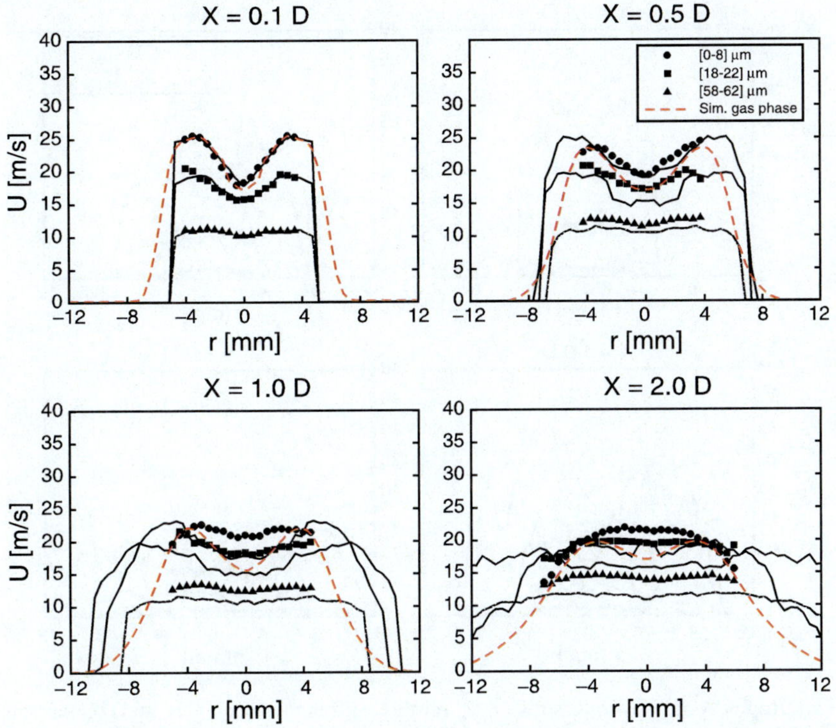

Fig. 5.9 Droplet axial velocity $\widetilde{U}_{\mathrm{p}}$ for three different size classes, and comparison with gas phase velocity

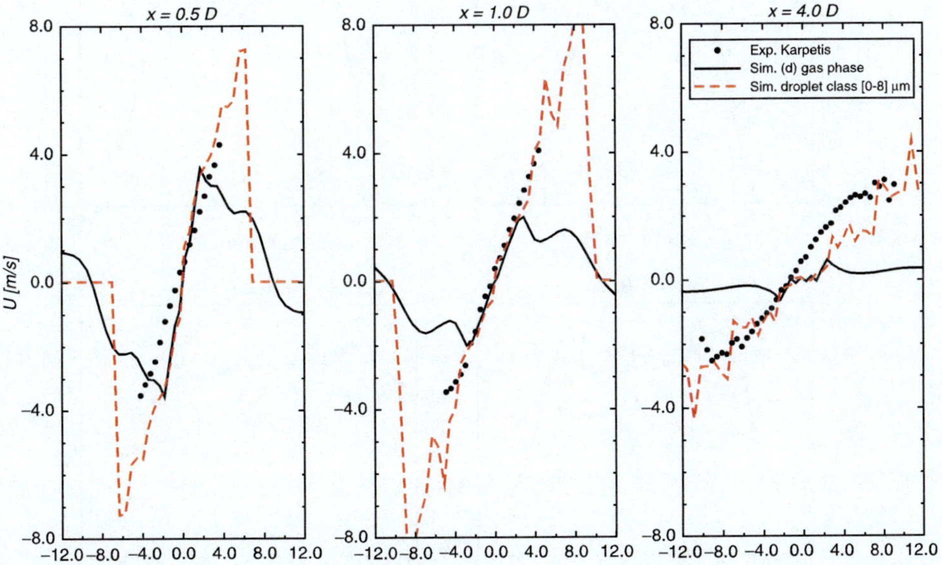

Fig. 5.10 Droplet radial velocity $\widetilde{V}_{\mathrm{f}}$. Comparison between gas phase and droplet radial velocity

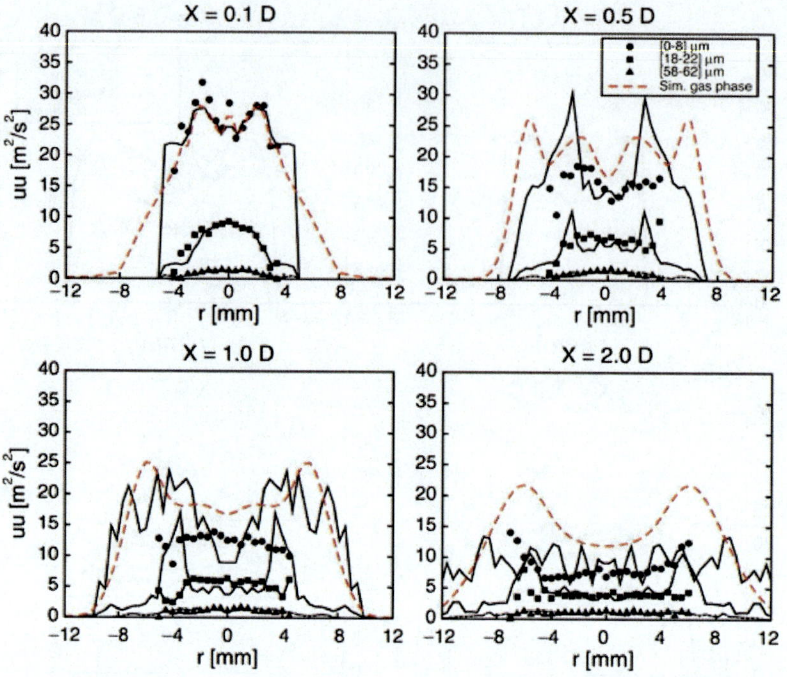

Fig. 5.11 Droplet Axial Reynolds stress $\widetilde{u_p'' u_p''}$ for three different size classes, and comparison with gas phase Reynolds stress

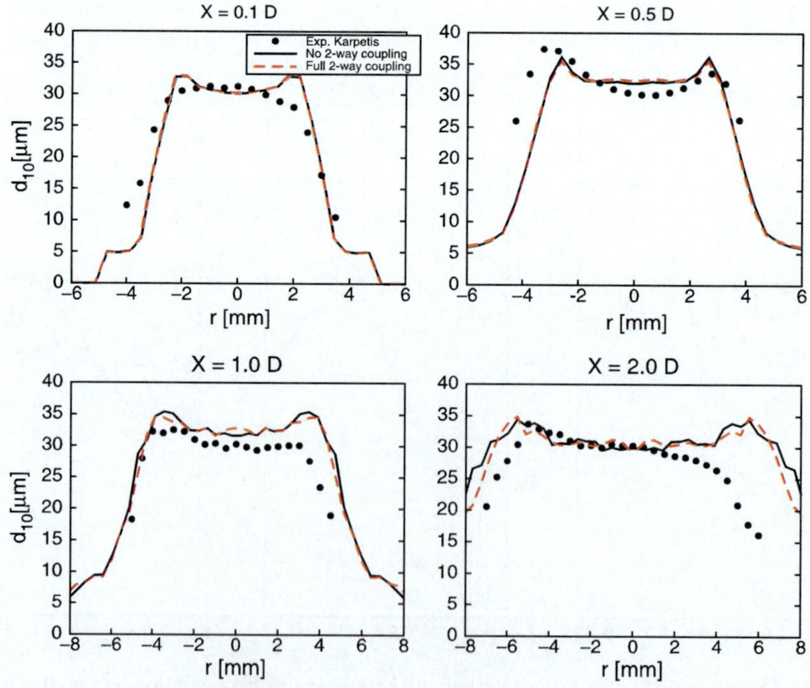

Fig. 5.12 Mean arithmetic droplet diameter d_{10} with and without two-way coupling

(finite volume approach) with a transported PDF approach (particle approach). The turbulence model for the pressure-rate-of-strain was extended for vaporizing dispersed two phase flows and a consistent Langevin model was derived. This turbulence model takes into account the effect of two-way coupling due to the presence of particles and the effect of mass transfer by vaporization. The new model was applied to a turbulent methanol spray flame studied experimentally by Karpetis and Gomez. The predictions of the two-way coupling models were compared to simulations without two-way coupling and to the experimental data. In general the observed differences between different treatments of the two-way coupling are small. The classical two-way coupling model already gives good predictions of the mean flow quantities. An important observation is that in the simulations, the smallest particles have no time to relax to the gas phase conditions due to vaporization.

Acknowledgements This projects was financially supported by the Dutch Technology Foundation STW

References

1. Lain, S., Sommerfeld, M.: Turbulence modulation in dispersed two-phase flow laden with solids from a Lagrangian perspective. Int. J. Heat Fluid Flow **24**, 616–625 (2003)
2. Gouesbet, G., Berlemont, A.: Eulerian and Lagrangian approaches for predicting the behaviour of discrete particles in turbulent flows. Prog. Energy Combust. Sci. **25**, 133–159 (1999)
3. Yuen, M., Chen, L.: On drag of evaporating liquid droplets. Combust. Sci. Technol. **14**, 147–154 (1976)
4. Drew, D.: Mathematical modeling of two-phase flow. Annu. Rev. Fluid Mech. **15**, 261–291 (1983)
5. Kataoka, I.: Local instant formulation of two-phase flow. Int. J. Multiphase Flow **12**, 745–758 (1986)
6. Naud, B.: PDF modeling of turbulent sprays and flames using a particle stochastic approach. Ph.D. thesis, Delft University of Technology (2003). http://www.darenet.nl/promiseofscience
7. Pope, S.: A Lagrangian two-time probability density function equation for inhomogeneous turbulent flows. Phys. Fluids **26**(12), 3448–3450 (1983)
8. Pope, S.: PDF methods for turbulent reacting flows. Prog. Energy Combust. Sci. **11**, 119–192 (1985)
9. Jenny, P., Pope, S., Muradoğlu, M., Caughey, D.: A hybrid algorithm for the joint PDF equation of turbulent reactive flows. J. Comput. Phys. **166**, 218–252 (2001)
10. Anand, G., Jenny, P.: Stochastic modeling of evaporating sprays within a consistent hybrid joint pdf framework. J. Comput. Phys. **228**, 2063–2081 (2009)
11. Pope, S.: On the relationship between stochastic Lagrangian models of turbulence and second-moment closures. Phys. Fluids **6**(2), 973–985 (1994)
12. Wouters, H., Peeters, T., Roekaerts, D.: On the existence of a generalized Langevin model representation for second-moment closures. Phys. Fluids **8**, 1702–1704 (1996)
13. Taulbee, D., Mashayek, F., Barré, C.: Simulation and Reynolds stress modeling of particle-laden turbulent shear flows. Int. J. Heat Fluid Flow **20**, 368–373 (1999)
14. Mashayek, F., Taulbee, D.: Turbulent gas-solid flows part I: direct simulations and Reynolds stress closures. Numer. Heat Transfer B **41**, 1–29 (2002)
15. Launder, B., Reece, G., Rodi, W.: Progress in the development of a Reynolds-stress turbulence closure. J. Fluid Mech. **68**, 537–566 (1975)

16. Speziale, C., Sarkar, S., Gatski, T.: Modelling the pressure-strain correlation of turbulence: an invariant dynamical systems approach. J. Fluid Mech. **227**, 245–272 (1991)
17. Beishuizen, N., Roekaerts, D., Naud, B.: Evaluation of a modified Reynolds stress model for turbulent dispersed two-phase flows including two-way coupling. Flow Turbulence Combust. **79**(3), 321–341 (2007)
18. Karpetis, A.: Self-similarity, momentum scaling and Reynolds stress in non-premixed turbulent spray flames. Ph.D. thesis, Yale University (1998)
19. Karpetis, A., Gomez, A.: An experimental investigation of non-premixed turbulent spray flames and their self similar behavior. In: 27th Symposium (International) on Combustion, pp. 2001–2008 (1998)
20. Daly, B., Harlow, F.: Transport equations in turbulence. Phys. Fluids **13**, 2634–2649 (1970)
21. Crowe, C.: On models for turbulence modulation in fluid-particle flows. Int. J. Multiphase Flow **26**, 719–727 (2000)
22. Beishuizen, N.: PDF modelling and particle-turbulence interaction of turbulent spray flames. Ph.D. thesis, Delft University of Technology (2008). http://www.darenet.nl/promiseofscience
23. Kolmogorov, A.: The local structure of turbulence in incompressible viscous fluid for very large Reynolds numbers. Proc. R. Soc. London Ser. A **434**, 9–13 (1991)
24. Pope, S.: Turbulent Flows. Cambridge University Press, Cambridge (2000)
25. Taweel, A., Landau, J.: Turbulence modulation in two-phase jets. Int. J. Multiphase Flow **3**, 341–351 (1977)
26. Druzhinin, O., Elghobashi, S.: On the decay rate of isotropic turbulence laden with microparticles. Phys. Fluids **11**(3), 602–610 (1999)
27. Druzhinin, O.: The influence of particle inertia on the two-way coupling and modification of isotropic turbulence by microparticles. Phys. Fluids **13**(12), 3738–3755 (2001)
28. Elghobashi, S., Truesdell, G.: On the two-way interaction between homogeneous turbulence and dispersed solid particles I: turbulence modification. Phys. Fluids A **5**, 1790–1801 (1993)
29. Squires, K., Eaton, J.: Measurements of particle dispersion obtained from direct numerical simulations of isotropic turbulence. J. Fluid Mech. **226**, 1–35 (1991)
30. Sundaram, S., Collins, L.: A numerical study of the modulation of isotropic turbulence by suspended particles. J. Fluid Mech. **379**, 105–143 (1999)
31. Hetsroni, G.: Particles-turbulence interaction. Int. J. Multiphase Flow **15(5)**, 735–746 (1989)
32. Wax, N.: Noise and Stochastic Processes. Dover, New York (1954)
33. Haworth, D., Pope, S.: A generalized Langevin model for turbulent flows. Phys. Fluids **29**, 1026–1044 (1986)
34. Xu, J., Pope, S.: Sources of bias in particle-mesh methods for PDF models for turbulent flows. Technical report, Cornell University, Ithaca (1997)
35. Xu, J., Pope, S.: Assessment of numerical accuracy of PDF/Monte Carlo methods for turbulent reacting flows. J. Comput. Phys. **152**, 192–230 (1999)
36. Minier, J., Peirano, E.: The PDF approach to turbulent polydispersed two-phase flows. Phys. Rep. **352**, 1–214 (2001)
37. Demoulin, F., Borghi, R.: Assumed PDF modeling of turbulent spray combustion. Combust. Sci. Technol. **158**, 249–271 (2000)
38. Fox, R.: Computational Models for Turbulent Reacting Flows. Cambridge University Press, Cambridge (2003)
39. Ge, H.W., Gutheil, E.: Simulation of a turbulent spray flame using coupled pdf gas phase and spray flamelet modeling. Combust. Flame **153**, 173–185 (2008)
40. Maxey, M., Riley, J.: Equation of motion for a small rigid sphere in a nonuniform flow. Phys. Fluids **26**, 883–889 (1983)
41. Simonin, O.: Statistical and continuum modeling of turbulent reactive flows – part I. Lecture Series 2000–06: Theoretical and Experimental Modelling of Particulate Flow, von Karman Institute, Fluid Dyn, B-1640 Rhode St Genese, Belgium (2000)
42. Bellan, J., Harstad, K.: Analysis of the convective evaporation of nondilute clusters of drops. Int. J. Heat Mass Transf. **30**, 125–136 (1987)

43. Miller, R., Harstad, K., Bellan, J.: Evaluation of equilibrium and non-equilibrium evaporation models for many-droplet gas-liquid flow simulations. Int. J. Multiphase Flow **24**, 1025–1055 (1998)
44. Chiang, C., Raju, M., Sirignano, W.: Numerical analysis of convecting, vaporizing fuel droplet with variable properties. Int. J. Heat Mass Transf. **35**, 1307–1324 (1992)
45. Karpetis, A., Gomez, A.: An experimental study of well-defined turbulent nonpremixed spray flames. Combust. Flame **121**, 1–23 (2000)
46. Muradoğlu, M., Pope, S.: A local time-stepping algorithm for solving the probability density function turbulence model equations. AIAA J. **40**, 1755–1763 (2002)
47. Minier, J., Peirano, E., Chibbaro, S.: Weak first- and second-order numerical schemes for stochastic differential equations appearing in Lagrangian two-phase flow modeling. Monte Carlo Methods Appl. **9**(2), 93–133 (2003)
48. Haworth, D.C.: Progress in probability density function methods for turbulent reacting flows. Prog. Energy Combust. Sci. **36**, 168–259 (2010)
49. Naud, B., Jiménez, C., Roekaerts, D.: A consistent hybrid PDF method: implementation details and application to the simulation of a bluff-body stabilised flame. Prog. Comp. Fluid Dyn. **6**, 147–157 (2006)
50. Naud, B., Merci, B., Roekaerts, D.: Generalized langevin model in correspondence with a chosen standard scalar-flux second-moment closure. Flow Turbulance Combust. **85**, 363–382 (2010)
51. Anand, G., Jenny, P.: A unified PDF formulation to model turbulence modification in two-phase flow. Phys. Fluids **21**, 105101–105117 (2009)

Index

B. Merci et al. (eds.), *Experiments and Numerical Simulations of Diluted Spray Turbulent Combustion*, ERCOFTAC Series 17, DOI 10.1007/978-94-007-1409-0,

T

U

V

W